Laboratory Manual for

Chemistry in Context

Applying Chemistry to Society

Ninth Edition

Edited by

Jennifer A. Tripp
University College London

and

Lallie L. McKenzie
Chem11, LLC

ACS
Chemistry for Life®

Mc
Graw
Hill
Education

LABORATORY MANUAL FOR

CHEMISTRY IN CONTEXT: APPLYING CHEMISTRY TO SOCIETY, NINTH EDITION

Published by McGraw-Hill Education, 2 Penn Plaza, New York, NY 10121. Copyright © 2018 by McGraw-Hill Education. All rights reserved. Printed in the United States of America. No part of this publication may be reproduced or distributed in any form or by any means, or stored in a database or retrieval system, without the prior written consent of McGraw-Hill Education, including, but not limited to, in any network or other electronic storage or transmission, or broadcast for distance learning.

Some ancillaries, including electronic and print components, may not be available to customers outside the United States.

This book is printed on acid-free paper.

1 2 3 4 5 6 QVS 21 20 19 18 17

ISBN 978-1-259-92013-4
MHID 1-259-92013-5

All credits appearing on page or at the end of the book are considered to be an extension of the copyright page.

The Internet addresses listed in the text were accurate at the time of publication. The inclusion of a website does not indicate an endorsement by the authors or McGraw-Hill Education, and McGraw-Hill Education does not guarantee the accuracy of the information presented at these sites.

mheducation.com/highered

Table of Contents

To the Instructor

This laboratory manual accompanies the ninth edition of *Chemistry in Context: Applying Chemistry to Society.* Both the text and this laboratory manual are designed for college students majoring in disciplines outside of the natural and physical sciences. *Chemistry in Context* examines a set of scientific and technological topics with broad societal implications; this manual provides laboratory experiences that are relevant to these topics. As authors and instructors, we believe that laboratory work ought to be an integral part of chemistry courses. Hands-on experience with investigations and data collection are crucial to an understanding of the scientific method and to the role that science plays in addressing societal issues.

For the prior eighth edition, each lab investigation underwent a major revision to utilize the Science Writing Heuristic (SWH), and that has carried over into the current edition. We encourage you and your students to read the introduction to SWH in this volume. With SWH, the students ask questions that can be answered experimentally, and then interpret and analyze their data and make claims about their results. Students are encouraged to write about the investigation and devise their own data tables and charts. As a result, no data sheets are included with the investigations. Nevertheless, we realize that many instructors find data sheets useful and we have made them available online along with the instructor's notes. They can be modified to suit the course and printed as desired.

Eight brand-new investigations have been added to this edition, including an investigation of sunscreen effectiveness, a simple synthesis of mauve dye, and an analysis of various substances to test for drugs. Other investigations have been significantly rewritten to reflect new topics presented in the textbook. We would like to thank Kevin Dunn (Hampden-Sydney College), for assistance in developing the mauve lab; Tom Davis (Claremont McKenna College), Roger Leonard, Clive Kittridge, and Craig Biersdorff (University of Oregon), for help with testing the new investigations; Michael Cann (University of Scranton), who wrote the introduction to green chemistry; Michael Mury, who advised us on the Science Writing Heuristic; and Emily Bones, who assisted in editing this new edition.

This collection of investigations is the result of a collaborative effort involving many people. Major contributions to earlier editions came from Catherine Middlecamp, Norbert Pienta, Truman Schwartz, Robert Silberman, Conrad Stanitski, Gail Steehler, and Wilmer Stratton.

All of the investigations in this lab manual have been used by *Chemistry in Context* authors, past and present, in our own classes. We invite users to contact us with suggestions for modifications of these investigations and/or for exchange of ideas about possible new investigations.

To the Student

The investigations in this laboratory manual have been carefully chosen and designed to reflect and amplify the topics in *Chemistry in Context.* Our goal in writing this manual was not to train you as a future chemist, but rather to illustrate what chemists do and how they do it. We also hope that you will appreciate that there is no great mystery to doing chemistry in the laboratory. You can obtain a great deal of information about the world around you with simple chemical equipment and straightforward procedures.

Another goal is to give you an opportunity to discover how chemists solve problems and to solve some yourself. After all, chemistry is a science of investigation and most chemists spend at least part of their time in a laboratory. You will use the laboratory to try out new ideas, investigate the properties of materials and compounds, synthesize compounds, analyze materials, and, in general, solve problems. Investigations are the way that scientists answer questions.

The investigations in this laboratory manual are in the format of the Science Writing Heuristic, a method that encourages you to mindfully develop questions and answer them through investigation. We encourage you to read the short introduction to the Science Writing Heuristic so that you can get the most out of the laboratory component of this course.

You will find that most of these investigations use simple equipment and that you can easily learn the necessary techniques. In general, you will be working with a partner, and in some cases the whole class will work collaboratively to collect data and answer a scientific question. Your instructor may ask you to keep a laboratory notebook or to fill in data sheets with your observations and results. In either case, it is important to keep good records about what you do in the laboratory.

We have been careful, when possible, to develop investigations that use minimally toxic reagents, and to use them in small quantities. You will be encouraged to consider the investigations in the context of **green chemistry**, a way of doing chemistry that seeks to protect the health of humans, wildlife, and the wider environment. Green chemistry is part of a greater movement toward *sustainability*, ideas of which you may have encountered in your other courses. We hope that the investigations in this book, together with ideas presented in the textbook, will enhance your understanding of some of the ways that chemists are helping to solve global problems.

Although you may never again work in a chemistry laboratory, it is our hope that after this laboratory course you will understand why chemists find laboratory work so interesting and compelling.

Green Chemistry

Ask any group of chemists what the world of chemistry has contributed to society and they will come up with a long list. The list might include: pharmaceutical drugs that help us live longer, healthier lives and ease our pain and suffering; polymers that make possible dashboards, skateboards, snowboards, and circuit boards; pesticides and fertilizers that without which we would have an even harder time of feeding the 7.5 billion hungry mouths on this planet; flavors and fragrances that we take for granted; cosmetics and personal care products that we use every day; batteries that power our flashlights, toys, cell phones, and laptops as well as start (and even power some of) our cars; and the list goes on and on.

Yet when non-chemists (such as yourselves) are asked what comes to mind when they hear the word "chemical," reciting these wonders of the modern world are generally not how they respond. More often they cite the toxic waste dumps that are down the block or across town from where they live, or depletion of the ozone layer by CFCs, or decimation of the bald eagle population by DDT, or other unpleasant effects of the chemical industry. Chemists, for too long, have not paid enough attention to the environmental consequences of the products they produce or the processes by which these products are made.

HOWEVER, there is a new paradigm "in town" and it is called **green chemistry**, chemistry with the environment in mind. Within the textbook that accompanies this lab manual, you will find many examples of green chemistry and discussion of how chemistry contributes to sustainability. As you can see from the textbook and the key ideas in green chemistry printed on the next page, green chemistry seeks to reduce the ecological footprint of chemical products and the processes by which they are made. Many of the investigations in this manual have been designed with green chemistry in mind. For example, Investigation 9 involves extracting the essential oil from orange peels using liquid carbon dioxide, a benign solvent that can replace toxic and flammable hydrocarbon solvents. Investigation 7 allows you to compare the properties of refrigerant gases and analyze how CFCs became popular and how alternatives are evaluated. In Investigation 14 you will synthesize biodiesel from vegetable oil, which offers a renewable alternative to petroleum-based diesel fuel used by cars and trucks. In Investigation 36, you will synthesize mauve and learn a little about industrial chemistry and sustainability. Many more investigations in this lab manual use and discuss green chemistry, and we encourage you to think about the key ideas as you do the investigations.

Green chemistry, along with other green technologies, is an integral part of our journey to find a path to sustainable development. Humanity's combined ecological footprint exceeds the carrying capacity of Earth. In short, we are depleting the natural capital of Earth (her renewable resources) and producing waste faster than Earth is able to convert waste back to natural capital. Green technologies not only offer the possibility of reducing the amount of resources that we consume, but even better, they offer an opportunity to convert waste back into natural capital. That is, technology becomes part of the solution, not part of the problem. We hope that by learning about green chemistry and sustainability, you are inspired to develop ways, in both your personal and (future) professional life, that you can contribute to sustainable development.

Michael C. Cann
University of Scranton

Key Ideas of Green Chemistry

1. It is better to **prevent waste** than to treat or clean up waste after it is formed.
2. It is better to **minimize the amount of materials used** in the production of a product.
3. It is better to **use and generate substances that are not toxic**.
4. It is better to **use less energy**.
5. It is better to **use renewable materials** when it makes technical and economic sense.
6. It is better to **design materials that degrade** into innocuous products **at the end of their useful life**.

The Science Writing Heuristic

If you ask your classmates and even your professors for the definition of inquiry, you will probably get as many unique definitions as the people you ask. Inquiry takes on different meanings for different people. In this lab manual, we focus on inquiry as the means for performing science. In order to have you think through the processes of science, we use headings and structure based on the Science Writing Heuristic (SWH).

SWH is a well-developed approach to guided inquiry experiences, and it is designed to encourage construction of conceptual knowledge. It is also based on relationships among questions, evidence, and claims. The traditional Science Writing Heuristic includes the following categories for students to process:

1. Beginning Questions—What are my questions?
2. Tests—What did I do?
3. Observations—What did I see?
4. Claims—What can I claim?
5. Evidence—How do I know? Why am I making these claims?
6. Reading—How do my ideas compare with other ideas?
7. Reflection—How have my ideas changed?
8. Writing—What is the best explanation that clarifies what I have learned?

We have adapted these categories for you into the headings below that you will see in each investigation.

Asking Questions

Scientific investigations usually begin with a question to be answered by gathering data and experimenting. Sometimes this question will be provided, while other times you will be asked to develop the question with your laboratory partners or classmates.

Preparing to Investigate

Before beginning experiments, it is important to clearly outline a procedure for gathering evidence that includes identifying the data to be collected and the steps to be followed. In some cases, a complete or partial procedure will be included in the investigation, but many times you will devise part or all of the procedure with your laboratory partners or classmates. Whether the procedure is provided or devised, you will need to study it completely before beginning. You will also need to create a system—usually a data table—for recording the observations and measurements you will make during the investigation.

Making Predictions

In some investigations, you will predict what you think will happen as you gather evidence. These predictions should be based on your prior experience and will not be evaluated for correctness, but you may be asked to reflect upon them after the investigation.

Gathering Evidence

The core of an investigation is gathering evidence. It contains directions, steps, or guidance for collecting data and observations.

Analyzing Evidence

The data gathered (evidence) in some investigations requires further processing before the initial question can be answered. Guidance is often provided to facilitate calculations or other analysis.

Interpreting Evidence

After analyzing evidence, you should ask, "What does the evidence mean?" Answering this question allows you to propose explanations for scientific phenomena. Questions within this section are designed to help you think about implications of the evidence and connect it to the purpose of the investigation.

Making Claims

Once data have been analyzed and interpreted, an answer to the initial question can be proposed. This answer often comes in the form of a scientific claim. Such claims must be supported by evidence from the investigation.

Reflecting on the Investigation

The final task in most investigations is to reflect on what was done, think about how your understanding has developed, and apply what was determined to other situations.

We hope these categories will assist you as you work through these investigations as well as help you think like a scientist.

Michael Mury
All Saints Academy, FL

Source: Greenbowe, T.J.; Hand, B. Introduction to the Science Writing Heuristic. *Chemists' Guide to Effective Teaching,* First edition; Pienta, N.J., Cooper, M.M., Greenbowe, T.J., eds. Prentice Hall, 2005.

Safety in the Laboratory

Good laboratory practice requires you take some simple safety precautions whenever you work in a chemistry laboratory. The popular notion that a chemistry laboratory is a dangerous place, filled with unknown disasters waiting to occur, is simply untrue for most situations, and is certainly incorrect for the activities in this laboratory manual. Nevertheless, all chemistry laboratories have some hazards associated with chemical spills, careless handling of flammable substances, and broken glassware.

With these in mind, here are some basic rules to follow when working in a chemistry laboratory.

1. **Protect your eyes with approved goggles or glasses.** This rule is essential and will be rigidly enforced. Failure to wear approved eyewear will result in dismissal from the laboratory. Chemical splashes can harm your eyes, and even dilute solutions of many chemicals can cause serious eye damage. You must wear eye protection even when you are not working directly with chemicals. Someone near you may have an accident and something may splash or fly in your direction.

2. **Exercise special care when using flammable substances.** Tie back long hair and avoid wearing clothes with loose sleeves. No open flames should be used anywhere in the laboratory, unless the experiment specifically calls for the use of a burner. Even a hot object can sometimes cause flammable vapors to ignite, so it is important for you to know which liquids are flammable.

3. **Never eat, drink, or smoke in the laboratory.** You may inadvertently ingest or inhale hazardous chemicals. It is also a good idea to wash your hands before exiting the laboratory.

4. **Never perform unauthorized experiments.** Some simple chemicals can form explosive or toxic products when mixed in unintended or inappropriate ways.

5. **Never work in a laboratory without proper supervision.** One of your best safety precautions is to have a knowledgeable person present who can spot potential hazards and handle an emergency should it arise. Notify your instructor immediately in case of any spill or injury, no matter how small. If you have any questions or concerns about safety, please address those with your instructor before proceeding with an experiment.

6. **Handle glassware carefully.** Glassware can break and cause nasty cuts. Report broken glassware to your instructor and dispose of it properly.

7. **Learn the location of safety equipment in your lab, including fire extinguishers, fire blanket, first-aid kit, eyewashes, and safety showers.** Be sure you know how and when they are to be used.

 STOP! Important safety information in each experiment is marked with this icon. Pay close attention to these cautionary notes and any additional safety information given by your instructor. Develop a habit of being safety-conscious whenever you are in the laboratory.

Lastly, make sure you always dispose of chemical waste in the appropriate waste container in your hood. Unless your instructor specifically tells you to do so, you should never put any of the substances used in the laboratory down the sink drain. Chemical waste must always be dealt with in a safe manner.

Laboratory Methods

In this laboratory course, you will be exploring the physical and chemical properties of matter. Because no one can directly see molecules, we will use a number of indirect techniques, some of which may be unfamiliar to you. This section introduces and explains the most common methods and procedures that you will use in the chemistry laboratory.

Table of Contents

Measuring Physical Properties

Much of what you will be doing in the laboratory involves taking measurements of properties such as mass, volume, and temperature—all physical properties of a substance. Here, we will go over a few basic techniques for measuring each of these properties. Keep in mind that actual laboratory equipment can vary and you should pay close attention to the instructions given by your instructor for the safe use and operation of the equipment available in your laboratory.

Measuring mass

The **mass** of a substance is a measure of how much matter it contains. Typically, mass is measured using an analytical balance, such as that pictured in Figure 0.1. The balance has a plate or pan to hold your sample, a display to read the mass, and often a clear box or dome that goes over the sample to prevent air currents from disrupting the measurement. In addition, there will be some buttons near the display.

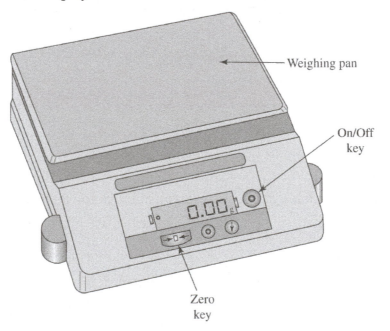

Figure 0.1. An image of an analytical balance.

Investigations in this laboratory will require a balance with accuracy to at least 1/100 of a gram (0.01 g, or 10 mg). To measure the mass of a sample, it is important to make sure that your balance reads exactly zero before making a measurement. Make sure that the balance pan is empty and press the "tare" or "zero" button. Your instructor will show you how to do this with your specific balance. Only after the display reads exactly zero should you put your sample on the pan. Allow the instrument to settle and write down the mass of your sample. Be sure to record the entire mass, even if the last digit is zero.

Measuring volume

Beakers, Erlenmeyer flasks, and other laboratory glassware have volume markings, but these markings are often not very accurate. The best way to measure the **volume** of a liquid sample is with a graduated cylinder (Figure 0.2). This piece of glassware has evenly spaced markings along the side of the glass cylinder, enabling you to read the volume with great accuracy. To measure the volume of a liquid, place the sample inside the cylinder and place the cylinder on a flat surface, such as the lab bench. Place yourself at eye level with the top of the liquid. You will see that most liquids are not flat on top, but rather have a curved surface called a **meniscus**. You should read the volume at the *bottom* of the meniscus, as shown in Figure 0.2. Read the value and then estimate the position between the two lines. For instance, the volume in Figure 0.2 should be read as 25.7 mL.

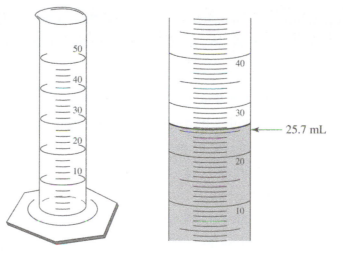

Figure 0.2. The volume of a liquid in a graduated cylinder should be read at the bottom of the meniscus.

Another common method for measuring volume is with a pipet. A pipet can be a tapered glass tube that uses a separate rubber bulb to draw the liquid into the tube, or an all-in-one plastic device as shown in Figure 0.3. These pipets come in a variety of sizes and may be graduated to show a range of volumes. To use a pipet, squeeze the bulb to push air out of the pipet, then, holding the pipet vertically with the bulb up, insert the tip of the pipet into the liquid and gently

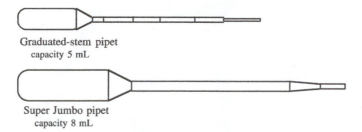

Graduated-stem pipet
capacity 5 mL

Super Jumbo pipet
capacity 8 mL

Figure 0.3. Examples of plastic pipets.

release pressure on the bulb until the desired volume of liquid is drawn into the tube. If you are using a glass pipet, take care that the liquid does not enter the bulb.

Measuring temperature

The **temperature** of a sample is a measure of how quickly the molecules in the sample are moving, on average. When a sample contains more heat energy, the molecules move more quickly and the temperature is higher. Several of the investigations in this book require the accurate measurement of temperature, and to do this you will use a **thermometer**. You will most likely be using an analog thermometer, a thin glass graduated tube filled with alcohol that has been dyed red. To measure the temperature, insert the bulb of the thermometer into the sample, and allow the red alcohol to come to a stable position. Read the temperature using the marked lines, and then estimate the position between the final two lines for additional accuracy.

Measuring temperature is even easier with a digital thermometer. If one is available, you simply insert the probe into the sample and read the temperature on the digital display.

Measuring pressure

Pressure is the amount of force that a substance, often a gas, exerts on its surroundings. Atmospheric pressure is measured with a **barometer**, which traditionally consists of a column of mercury connected to a bulb. The column of mercury moves up and down depending on the ambient air pressure. Atmospheric pressure at sea level is typically 760 mmHg or Torr, but varies with elevation and weather patterns. More modern instruments for measuring air pressure do not involve mercury, a toxic metal. Should you be required to measure atmospheric pressure, your instructor will provide instructions about how do it.

Pressure can differ significantly from atmospheric pressure in enclosed systems such as tires or soda bottles. If you are required to measure the pressure inside a closed system, you will use a pressure gauge, such as that used to check the air pressure in your car or bicycle tires. These come in many varieties (Figure 0.4) and your instructor will demonstrate the use of the ones you have available.

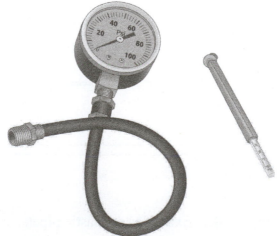

Figure 0.4. Examples of tire pressure gauges.

Additional measuring methods

Some investigations in this book require you to measure other quantities, such as the conductivity, density, or melting point of a substance. Details about these measurements will be given in the individual lab procedures.

Analyzing Chemical Properties

In contrast to physical properties, chemical properties are those that result from molecular-level change within a substance. Properties such as pH, heat of combustion, and chromatographic affinity are useful for studying chemical reactions and analyzing the chemical composition of a substance. Several instrumental and **wet chemistry** methods will be used in this course.

Measuring pH

Indicator paper

You will learn during this chemistry course that certain substances can act as **acids** and others as **bases. pH**, described in detail in Chapter 8 of the textbook, provides an indication of the presence or absence of acids, as well as their strength and concentration. A number of ways to measure pH have been developed. The simplest way to measure acid/base properties is by using litmus paper, which turns red in an acid and blue in a base. More sophisticated pH papers have been developed that use more than one indicator dye to give different color changes through the pH range (0 to 14).

Digital pH meter

You may also measure pH with a digital pH meter (Figure 0.5). This consists of a small box with a digital display and several buttons or knobs, connected by a cable to a probe called a pH electrode. Some instruments must be plugged into a wall socket, while others run on batteries. Note that the probes are rather fragile, expensive, and to be treated with care. The probe should be rinsed and wiped carefully and never allowed to dry out. The pH meter must be calibrated before use, and to do this you will use standard buffer solutions of known pH.

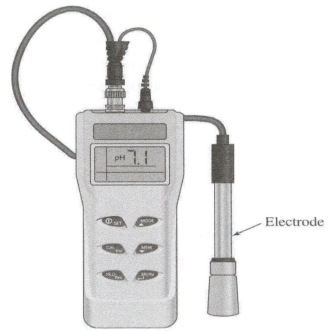

Electrode

Figure 0.5. Digital pH meter.

Follow these instructions for calibrating the meter and using it to measure the pH of a sample:

1. Obtain two small beakers containing the standard buffer solution. These usually come in pH 4.00, 7.00, and 10.00. Your instructor will indicate which to use for your investigation.

2. Hold the electrode over a waste container (usually a beaker) and rinse the electrode thoroughly with pure water from a wash bottle. Blot the end of the electrode *gently* with a soft tissue to remove most of the water.

3. Insert the electrode into the higher pH buffer solution, stir gently for a few moments, and observe the reading on the meter. Use the "calibrate" knob or buttons to adjust the meter until it displays the correct pH, within 0.1 pH units (*i.e.*, for a pH 10.00 reference solution, the calibrated meter should read between 9.9 and 10.1).

4. Again, rinse the electrode thoroughly with pure water, blot the end gently with a tissue, and then insert the electrode into the lower pH calibrating solution. Stir and observe until the reading is steady. Use the "slope" or "temperature" control, as indicated by your instructor, until the display reads close to the correct pH.

5. Repeat the measurements with the two buffer solutions (rinsing and blotting each time the electrode is moved) to ensure that the readings are steady.

6. To measure a sample: Thoroughly rinse and gently blot the pH electrode. Insert the electrode into the sample and stir gently. Without touching any knobs or buttons write down the pH indicated on the display. Between samples, wash and blot the electrode. When the electrode is not in use, it must stay submerged in water or another solution as indicated by your instructor so that it does not dry out.

Other methods

A third method for measuring pH is titration, which uses a chemical indicator and volumetric analysis to determine the concentration of acid or base in a solution. Titration is described in detail later.

Use of indicators

Many chemical changes are invisible, so other ways to observe and study them must be applied. A simple way to determine whether a chemical change has taken place is to use an **indicator**. Indicators are chemical substances added to reactions to provide a visible signal, usually a change in color or formation of a solid precipitate, which demonstrates that another change has taken place. In this course you may use indicators that change with changes in pH, concentrations of various anions and cations, and vitamin C. In general, you will add a drop or two of the indicator solution to the reaction or sample, and then observe the appropriate change.

Titration Method of Analysis

The **titration** method of analysis is extremely useful for determining the presence and concentration of reactive molecules or ions in a solution. In a titration, a known volume of the solution containing the substance of interest is measured, then a solution of a compatible reactant is slowly added until a complete reaction is observed. For instance, if you wish to determine the amount of acid in a solution, you would carefully measure a quantity of the acid and then slowly add a base, making sure to measure the volume added, until you neutralize the solution. If the concentration of the solution of base is known, then you can calculate the concentration of the acid. To determine the endpoint of a titration, an appropriate indicator is added so that it can be observed.

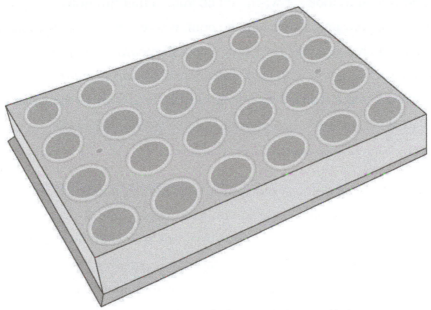

Figure 0.6. A reaction wellplate.

The titrations performed in this course will be done on a small scale, using a 24-well plastic wellplate (Figure 0.6) for the reactions and plastic transfer pipets to add the solutions to the wells. For colored indicators, you should place the wellplate on white paper to best observe the color change. The best way to observe formation of a solid precipitate is to place the wellplate on a black countertop or piece of paper. Be sure to clearly label all solutions and pipets so that your reagents do not get mixed up during the titration.

Rather than directly measure the volume of the solutions, we will use plastic pipets to drop solutions into the wellplate. This method requires assuming that every drop from the pipet holds the same volume. The first time you perform a titration, you should practice with the pipet.

Fill a pipet with water and dispense drops into a well of the wellplate. To successfully perform a titration, you will need to confidently add a known number of drops to the bottom of a well, one drop at a time. Squeeze the bulb gently, while the pipet is held vertically and directly over the center of the well. You may find it helpful to use two hands to steady the pipet. Plastic transfer pipets easily acquire air bubbles in their stems, which leads to frustrating partial drops that introduce errors. When you are doing a titration, you should reserve one well for solution waste. During a titration, these partial drops can be added to the waste well rather than the titration well. Once you can confidently add a known number of whole drops to a well, proceed to the titration.

To do a trial run, add 10 drops of your sample and a drop of indicator into a well, and then add your reactant solution dropwise until you observe the indicator change. You may need to stir the reaction with a small stirrer or toothpick for the change to persist. Continue adding a few more drops of the reaction solution to see the extent of the chemical change. By doing this trial run, you will know what to look for when you do your actual titrations.

In doing titrations, your goal is to catch the point where *one drop* of the reactant solution causes the first permanent change to the indicator. You will record the number of drops required to do the titration, and use this along with the known concentration of the reactant solution to calculate the unknown concentration of the sample. Further details about the procedures and the calculations will be given in the investigations that require titration analysis.

Spectroscopy

Spectroscopy is the study of how electromagnetic radiation interacts with matter. The full electromagnetic spectrum includes radio waves, microwaves, and X-rays, as well as infrared, visible, and ultraviolet light (see Chapter 3 in *Chemistry in Context* for more details). Each color and each unique position in the electromagnetic spectrum is identified by its **wavelength**. For example, the visible region contains light with wavelengths in the range of billionths of a meter, and, therefore, the wavelengths are expressed as **nanometers**, nm, (1 nm = 1×10^{-9} m).

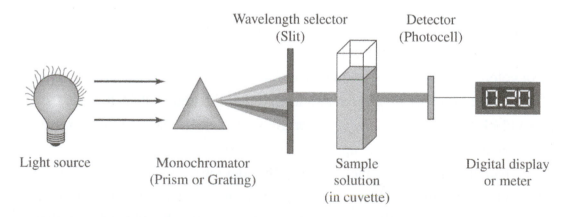

Figure 0.7. Simple diagram of how a spectrophotometer works.

In this laboratory course, you will measure how molecules absorb visible and ultraviolet (UV) light. The instrument used to measure the interaction of light with matter is a **spectrophotometer**. This device is designed to split visible light into its component colors (*i.e.*, different wavelengths) and then allow light of a selected wavelength region to pass through a sample of the material before being studied. An electronic detector measures the amount of light that has been transmitted or absorbed by the sample at each wavelength. All spectrophotometers have essentially the same components but vary in their degrees of sophistication. These essential components are (1) a light source, (2) a device to isolate or resolve particular wavelengths of light, (3) a sample holder, (4) a detector, and (5) a meter, a computer, or another device to display the measured transmittance or absorbance of light. Figure 0.7 shows a simple block diagram of spectrophotometer components.

Spectrophotometers are available for most regions of the electromagnetic spectrum. Spectrophotometers for the visible region of the spectrum were developed first and are still the most common. In these instruments, the light source is an ordinary incandescent light bulb; the wavelength selector consists of a diffraction grating and some lenses; a liquid sample is placed in a container resembling a test tube; and the detector is a phototube that converts light intensity into an electrical signal. Different investigations in this book require spectrophotometers that can measure absorbance in the visible and UV regions.

Simple spectrophotometers of the type described here can measure the transmittance of light of only a single wavelength at a time, but it's often useful to measure how a sample interacts with light at a number of different wavelengths. Obtaining data over a range of wavelengths requires multiple measurements, each made at a different wavelength. Because light intensity and detector sensitivity vary with wavelength, each measurement must be corrected using a **blank sample**, often pure water.

 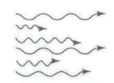

Incoming light: all wavelengths represented

Sample: absorbs some of the light

Outgoing light: only light that wasn't absorbed gets through

Figure 0.8. Absorption of light by a sample using spectroscopy.

The meter on the instrument you use will display either **absorbance (A)** or **percent transmittance (%T)** (Figure 0.8). The percent transmittance focuses our attention on the light that passes unaffected through the sample. Percent transmittance is defined as the ratio of light that passes through a sample to the light that passes through a blank of equal thickness, and multiplied by 100.

$$\%T = \frac{\text{transmittance by sample}}{\text{transmittance by blank}} \times 100\%$$

Absorbance, the other spectrophotometer unit, focuses our attention on the portion of light that gets absorbed by the sample. Absorbance is proportional to the concentration of the solution. A sample that absorbs no light will have 100% transmission and zero absorbance. Your instructor will let you know which unit to use for the investigations that involve spectroscopy.

When you make your measurements, you should use two identical test tubes (or other containers) to measure the blank and the sample. With the blank in place, you will adjust the spectrophotometer to read 100%T or zero absorbance. You will then insert your sample and record data for your sample by reading the displayed value. You must re-insert the blank and zero the instrument for each wavelength you measure. Once you measure %T or A over a range of wavelengths, you can plot the data with wavelength along the *x*-axis and your data on the *y*-axis. The result is a **spectrum**, a curved line that shows how a particular solution transmits or absorbs light in the visible region of the electromagnetic spectrum. More sophisticated spectrophotometers with computerized detectors may automatically create the spectrum for you.

Spectrophotometers are sensitive and expensive devices, so follow any specific guidelines from your instructor regarding the operation of instruments present in your laboratory.

Chromatography

The term **chromatography** comes from the Greek meaning "to write with color"; the first use of chromatography was to separate and study colored pigments in plants. In its simplest form, a chromatographic set-up consists of an immobilized solid, called the **stationary phase**, over which a gas or liquid, called the **mobile phase**, moves. A variety of chromatographic techniques have been developed, and all chromatographic techniques make use of the fact that components of a mixture injected into the mobile phase can be separated based on their tendency to move along with the mobile phase rather than be attracted to the stationary phase.

A simple method of chromatography involves using a piece of absorbent paper as the stationary phase and a liquid, such as water, as the mobile phase (Figure 0.9). Place small spots of your sample near one edge of the paper, and that edge is placed in the water. The liquid moves up the paper by capillary action.

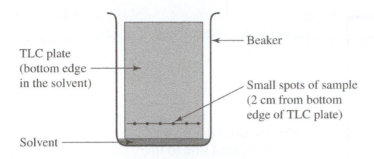

Figure 0.9. TLC set-up.

A slightly more sophisticated method is **thin-layer chromatography**, or TLC, in which the stationary phase consists of a glass, plastic, or aluminum plate coated in a suitable solid substance, such as silica or alumina. The mobile phase is a liquid, often a mixture of organic solvents that provide a particular polarity. As above, the sample is spotted on one end of the plate, and that edge immersed in the solvent so that the mobile phase moves up the plate by capillary action, carrying along the components of your sample mixture at different rates to separate them. This is commonly called "developing" the plate.

Whether the stationary phase is paper or a TLC plate, it is important to mark your sample spots with pencil, so that you know which samples you have analyzed. Multiple spots may be applied to a single plate or paper, but should be spaced about 1 cm apart. Depending on the volatility of the solvent used, it may be necessary to cover the beaker with a lid to prevent rapid evaporation. When the solvent front has reached a point about 1 cm from the top of the plate, remove the plate from the solvent, and immediately mark the position of the solvent front with a pencil.

If you are analyzing colored samples, you should be able to directly see the colors on the plate or paper. If not, some other method will be necessary to make your spots visible. One method is to expose the developed plate to a compound that will react chemically with the spots to make them visible. Another method is to examine the plate under UV light to see if any of the components are fluorescent. A related technique is to incorporate a fluorescent dye into the plate so that when it is examined under a UV lamp, the plate will glow (fluoresce) everywhere except where the component spots are.

Quantitative analysis of TLC plates is possible by calculating the R_f. R_f is defined as the distance traveled by the compound on the plate, divided by the distance traveled by the solvent. It should always be less than 1, since the compounds cannot move further along the plate than the solvent. You calculate R_f by first using a ruler to measure, in mm, the distance from the spot line to the front of the spot, and the distance from the spot line to the solvent front (Figure 0.10).

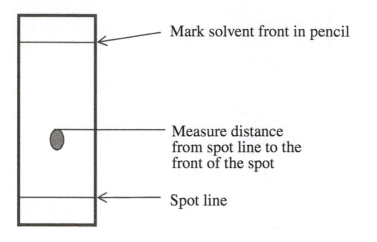

Figure 0.10. Calculating R_f on a TLC plate.

R_f is then easily calculated using the following equation.

$$R_f = \frac{\text{distance to spot}}{\text{distance to solvent front}}$$

Calorimetry

An important property of some materials is how much heat they release when burned, and the method for measuring this quantity is called **calorimetry**. In the method used in this course, the heat from combustion of a fuel will be used to heat a known volume of water. Accurate calorimetry depends on the accuracy of the mass measurements you make during the investigation, so it's a good idea to review the earlier section on measuring mass before beginning calorimetry. A diagram of the investigation set-up can be seen in Figure 0.11.

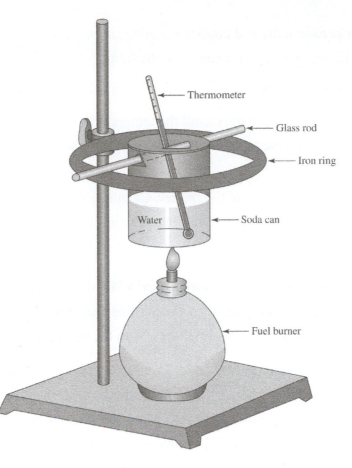

Figure 0.11. Diagram of can-and-burner calorimetry setup.

Follow these steps to set up the apparatus and perform a calorimetry investigation:

1. Obtain a dry soda can with the top removed and two holes punched on opposite sides near the top. Measure and record the mass of the empty can.

2. Add approximately 100 mL of water to the can. Measure and record the mass of the can with the water. Calculate the mass of water by subtracting the mass of the can.

3. Slide a glass rod through the holes in the can so that it can be suspended from the ring attached to a ring stand, as shown in the diagram.

4. Put a thermometer in the can, stir the water for a few moments, and then measure and record the temperature of the water.

5. Obtain a burner filled with fuel. Measure and record its mass.

6. Place the burner under the can, and adjust the height of the ring so that the bottom of the can is about 2 cm above the top of the wick.

7. Light the burner and observe the flame. If necessary, cautiously adjust the height of the can so that the top of the flame is just below the bottom of the can.

8. Stir the water occasionally and continue heating the water until the temperature has increased by about 20 °C; then extinguish the flame.

Quickly do the following two steps:

9. Continue stirring the water gently until the temperature stops rising; then record the temperature. Calculate the temperature change by subtracting the initial temperature from the final temperature.

10. Measure and record the mass of the burner. Calculate the mass of the fuel burned by subtracting the final mass of the burner from the initial mass.

Calculations

Using the data you have obtained, you can calculate the heat absorbed by the water. Theoretically, the amount of heat liberated by the burning fuel should equal the heat absorbed by the water, but in practice, some of the heat will be lost to the surroundings.

The **specific heat** of water is 1.00 cal/g•C, meaning that it takes exactly 1 calorie (cal) of heat to raise the temperature of 1 gram (g) of liquid water by 1 °C (see Chapter 5 in *Chemistry in Context*). Therefore, you can use the mass (m) of water and the amount of temperature change (ΔT) to calculate the total heat absorbed by the water.

$$\text{heat absorbed (cal)} = m(g) \times \Delta T(°C) \times 1.00 \text{ cal/g} \bullet °C$$

Confirm that the units on the right side of the equation cancel out, leaving only cal. Next, calculate the calories of heat per 1 g of fuel so that you can compare the heat values for different fuels based on their mass.

Running a Chemical Reaction

In this course, you will have several opportunities to transform matter by running a chemical reaction. Of course, most of the methods for analyzing chemical properties described above involve chemical reactions, but when a reaction is run on a large scale with the goal of isolating a product, some different techniques are required. Those are described here.

Heating a reaction

Often, one must heat a chemical reaction in order to speed it up. Still, other investigations involve heating a substance to measure its physical or chemical properties. You will use several different methods for heating a reaction, each chosen because of its suitability for the particular substance that needs to be heated.

Bunsen burners

A Bunsen burner is simply an apparatus to control the flame from burning natural gas. The burner has a diffuser at the top to assist with even heating, and the amount of air entering the burner, and thus the intensity of the flame, can be adjusted by adjusting a valve at the base of the burner.

An important safety note: Bunsen burners should *never* be used to heat flammable substances. Always ensure that no flammable chemicals are in the vicinity of the burner. When you use a Bunsen burner, you should tie back your hair and avoid wearing clothes with loose sleeves.

Bunsen burners are most often used to heat small amounts of non-flammable solutions, or to perform a flame test, in which a small amount of a solid is placed into the flame to see what color it burns. Heating a test tube over a flame must be done carefully, as shown in Figure 0.12. Heating the tube too rapidly, especially if it is held on an upright position, will cause the hot contents to splash out of the tube. In addition to being quite dangerous, it will require you to start over with a fresh sample.

To perform a flame test, a small amount of solid or liquid sample is placed on the end of a copper wire, and the wire is immersed in the flame. When the flame changes to different colors, it can indicate the presence

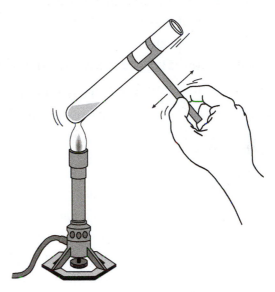

Figure 0.12. Heating a test tube over a Bunsen burner.

of certain elements. Further details about flame tests will be given in the investigations that require them.

Hot plates

Hot plates are an efficient way to heat large amounts of reagents, and are much safer to use with flammable substances. A good way to heat reactions to a specific temperature is to prepare a hot-water bath, by placing a partially full beaker of water on the hot plate and heating it to a particular temperature, measured by a thermometer. Your test tubes or other reaction containers can then be placed into the water bath to be heated.

Methods for purification of chemical substances

Measurements often require a pure chemical substance, so knowing some techniques for isolating one compound from a mixture will assist you in your measurements. If you are interested in studying compounds from nature, you often must separate the different components of the plant or food you wish to study. If you perform a chemical reaction, the desired product must be isolated from any solvents, by-products, or unreacted starting materials. Here are three purification methods.

Extraction

Extraction is used to separate substances of different polarities based on their solubility in water. Perhaps you have observed that some kinds of salad dressings separate—the water goes to the bottom of the bottle and the oil to the top. This separation occurs because water and oil are not **miscible**, meaning that they do not mix. Any **hydrophilic** (water loving) compounds in the salad dressing will dissolve in the water and sink to the bottom of the bottle with the water, while the **hydrophobic** (water fearing) compounds prefer to dissolve in the oil and stay at the top of the bottle with the oil. This is exactly the principle of extraction.

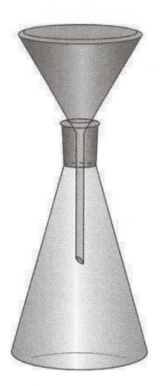

Two compounds can be separated if they are immiscible, or if they dissolve in two different solvents that are not miscible. More specific details about extraction procedures will be provided within the investigations that require this technique.

Filtration

Filtration is a simple method for separating a solid from a liquid. Perhaps you have used a colander to strain pasta after cooking; the colander acts as a filter to separate the solid pasta from the liquid cooking water. In the chemistry lab, we often use coarse paper as the filter, which is placed into a funnel and the mixture poured through. Any solids get caught in the filter paper, while the liquids flow through into a container.

Gravity filtration uses a filter funnel, a cone-shaped glass funnel with a narrow tube extending from the bottom, sometimes with ribs to increase filtering efficiency. Filter paper is folded to fit inside, and the funnel is placed over an Erlenmeyer flask (Figure 0.13). When the mixture is poured into the funnel, the solid collects in the filter paper while the liquid goes through into the flask.

Figure 0.13. Gravity filtration setup.

Recording Your Data

It is important, as you perform your investigations, to take adequate notes on what you do, measure, and observe. Carefully record your procedures, write down the measurements you make (including the units!), and any observations about your investigation. Write down your predictions about the investigation, the questions you seek to answer, and the results. Record your calculations and write conclusions. Your instructor may ask you to keep a laboratory notebook and create your own data tables, or give you data sheets to record your results for each investigation.

Data Analysis

Once you have recorded your measurements you will need to look carefully at your data and observations to draw conclusions. Sometimes you can do this by directly reviewing your results, but often you will need to further analyze your data by performing calculations or making graphs.

Calculations

Many investigations in this laboratory manual require calculations. The purpose of the calculations is usually to relate a quantity that you have measured (such as the volume of base used to titrate an acidic solution) to another quantity you are interested in but can't measure directly (such as the amount of acid in the solution). Details of calculations for each investigation will be given within the individual investigations. When doing calculations, make sure that the units of your numbers cancel out as they should, and always report the unit of your answer with the number in your lab reports.

Graphing

Large amounts of data are often best presented as a graph, a pictorial way of presenting information that will allow you to see trends and relationships between the data you collect. Figure 0.14 shows how data relating mass and volume can be shown as a graph.

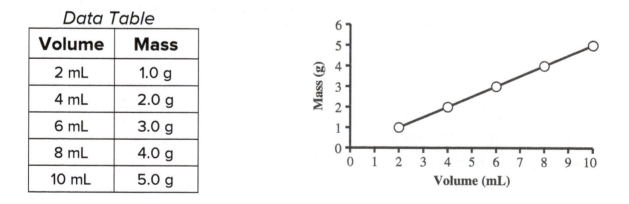

Data Table	
Volume	**Mass**
2 mL	1.0 g
4 mL	2.0 g
6 mL	3.0 g
8 mL	4.0 g
10 mL	5.0 g

Figure 0.14. Graphical representation of data: volume as a function of mass.

The graph in the figure shows that as the volume increases, the mass increases in direct proportion. Data that produce a straight line when plotted are said to have a linear relationship. Representing this data with a graph makes it straightforward to estimate the value of the mass for a volume that is in between measured data points, a process known as **interpolation**. For example, if the volume was 5 mL, from the graph, it's clear that its mass should be 2.5 g. One can also extend the line of the graph and obtain data beyond the range of measured points, a process known as **extrapolation**. Thus based on this data, if the volume were 15 mL the mass should be about 7.5 g. Care must be taken when extrapolating certain data, however, because the trends observed in one range of a graph are not necessarily sustained in other regions of the same graph.

A straight-line relationship, such as that in Figure 0.14, can be summarized with a simple algebraic equation of the form $y = mx + b$. This is called a **regression** line. In this equation, x and y are the values for the two quantities being plotted (*e.g.*, volume and mass), m is the **slope** of the plotted line, and b is the **y-intercept** (the value of y when the line crosses the y-axis, or the point where $x = 0$). The slope of a line can be calculated using the equation below. In this equation, $y_2 - y_1$ is the difference between the y values for two data points, and $x_2 - x_1$ is the difference between the x values for the same two data points.

$$\text{slope} = m = \frac{y_2 - y_1}{x_2 - x_1}$$

The slope summarizes the relationship between the columns of data. In the example above, it is easy to see that the slope is 0.5 g/mL, and it represents the relationship between mass and volume, or the density of the substance. Because the line crosses the y axis at 0, the y-intercept is 0, and the equation that represents the line is $y = 0.5x + 0$, or simply $y = 0.5x$.

While it is often useful to sketch out a graph by hand, using a computer graphing program takes much of the tedium out of the process. Many programs, for example, can create a complete graph from your data and generate a linear-regression line, using a statistical method to calculate the best straight-line fit for a set of data points. This can be particularly useful when there is a lot of "scatter" in the data, as in the graph in Figure 0.15.

Several investigations in this lab manual ask you to create graphs, which can be done by hand or on a computer as requested by your instructor.

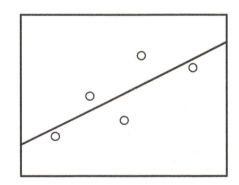

Figure 0.15. Linear regression fit of data points.

Preparation and Properties of Gases in Air

Asking Questions

- What is in the air you breathe?

- What gas makes up the majority of the air you breathe?

- Which gases are important to sustaining life?

- What are some of the sources of carbon dioxide in air?

- What are some differences between the air you inhale and the air you exhale?

Preparing to Investigate

In this investigation, you will prepare samples of two gases in the air, oxygen and carbon dioxide. You will then investigate some of the properties of the gases.

To prepare oxygen, you will use potassium iodide (KI) as a catalyst to decompose a familiar household product, hydrogen peroxide, into water and oxygen.

$$2 \ H_2O_2 \xrightarrow{\text{catalyst}} 2 \ H_2O + O_2$$

Carbon dioxide will be prepared from another common household product, baking soda, which has the chemical name sodium bicarbonate ($NaHCO_3$). When acetic acid (vinegar, $HC_2H_3O_2$) is mixed with sodium bicarbonate, a chemical reaction occurs, which forms a salt, sodium acetate ($NaC_2H_3O_2$), water (H_2O), and carbon dioxide (CO_2).

$$NaHCO_3 + HC_2H_3O_2 \longrightarrow NaC_2H_3O_2 + H_2O + CO_2$$

Both gases will be generated in zipper plastic bags, and you will have an opportunity to make observations about the reactions. Samples of the gases, as well as samples of air and exhaled air, will be tested for flammability, using a burning wood splint, and for reactivity, with an aqueous solution of calcium hydroxide, $Ca(OH)_2$, also known as limewater.

Finally, you will investigate what happens when these gases dissolve in water. In particular, you will determine whether or not they react with water to form an acid. Chapter 8 in your textbook explains concepts of acidity and pH, as well as some of the effects that rising atmospheric CO_2 levels are having on the oceans. To test for changes in acidity, you will use an

NOTES

Investigating Air Pollution

Asking Questions

- Explain why ozone at ground level is harmful, even though ozone in the upper atmosphere is essential to protect life on Earth. What are the primary effects of ground-level ozone on human health? On other living things?
- How is ground-level ozone formed? Identify at least three locations around your home or school that are likely to have higher concentrations of ozone.
- Particulate pollution is classified by its size. Identify the standard reported size values for particulate matter and rank them by their harm to human health.
- How is particulate pollution formed? Identify at least three locations around your home or school that are likely to have significant concentrations of particulates.
- What is a *variable* in an experiment? Define *dependent variable* and *independent variable*.

Preparing to Investigate

As you have read in Chapter 2 of *Chemistry in Context*, human activity can alter air quality. Two common pollutants are ozone and particulate matter. Ozone in the upper atmosphere is essential for blocking harmful ultraviolet (UV) light from reaching the surface of Earth, but ground-level ozone can reduce lung function when breathed by animals and can also adversely affect plant growth. Particulate matter is classified by size and usually reported as PM_{10}, which consists of particles with an average diameter of 10 μm, and $PM_{2.5}$, which have an average diameter of 2.5 μm. All of these particles can be inhaled deeply into the lungs and cause irritation. Additionally, the smallest particles can pass into the bloodstream, travel around the body, and damage other tissues, including the heart.

Ozone and particulate matter can form naturally, from sources including lightning, forest fires, and volcanoes. However, human activities involving combustion, such as driving a car, manufacturing goods, or generating electricity, can increase the concentrations of these pollutants in the air. In this experiment, you will prepare simple devices that will allow you to detect ozone and particulates and use them to measure how the concentration of these pollutants varies at different locations.

Making Predictions

Before beginning the experiment, you must choose your outdoor test locations. For each experiment, you should choose a control location where you expect to have very little pollution, and then three test sites where you expect pollution to be present. The locations do not need to be the same for the ozone experiment and the particulate experiment. For the ozone detection, your test sites must not be in direct sunlight. Each experiment should have an independent variable that changes between test sites, while other variables stay the same. When choosing your test sites, consider variables such as proximity to roads or factories, height above the ground, presence of vegetation, wind speed and direction, and time of day.

You can use different locations for the ozone experiment and the particulate experiment. What is the independent variable in each experiment? What other variables must you control? What variables can you not control? What variables are the same for all sites? Formulate a hypothesis that predicts how the pollution levels (ozone or particulates) with change in each location.

Gathering Evidence

Overview of the Investigation

1. Prepare ozone test strips and particulate test cards.

2. Mount in your test locations, plus a control location, and then analyze them.

Part I. Detecting Ozone

1. Measure 50 mL of distilled water and 4 g of corn starch. Add both to a 150-mL beaker. Gently heat the beaker while stirring with a glass rod or magnetic stirrer until the mixture becomes a thick and translucent gel.

2. Remove the beaker from the hot plate and add 1.5 g KI. Stir until well mixed, then let the mixture cool for two minutes.

3. Construct your four test strips by obtaining four pieces of filter paper at least 50 mm in diameter. Place them each on a large watch glass. Use a small paint brush to apply the gelled solution to both sides of each piece of filter paper. Make the layer of gel as uniform as possible.

4. Determine how you will label and mount your test strips, keeping in mind that they must hang freely.

5. Spray each strip with distilled water and install them at your three test sites and your control site. Leave each strip in place for 30 minutes.

6. To analyze a strip, spray it with distilled water and compare each strip with the control. Record your observations. If strips are not going to be analyzed immediately, store them in a sealed plastic bag or glass jar in a dark place.

Part II. Detecting Particulates

1. Obtain four index cards, one for each of your test sites and your control site. Cut a 3 cm × 3 cm square hole in the middle of each card.

2. Devise a method for mounting your test cards, and make any necessary modifications to the cards without altering the square hole. Label each card with the test location.

3. Place a piece of transparent packaging tape over each square hole. Avoid touching the sticky part of the tape within the square hole.

4. Mount the cards at the test sites with the sticky side facing outward. Leave the cards in place for at least one day, or up to one week. If rain or snow is likely, retrieve the cards and store them in plastic bags until the precipitation has passed.

5. When you collect the cards, take care not to touch the sticky side of the tape. Cover the sticky portion of the tape with a fresh piece of packaging tape to seal the particles between two pieces of tape.

6. Use a microscope to observe the particles. Devise a method for counting the particles and describe their size and properties. Record your data and observations.

Clean up

Dispose of your test strips in an appropriate waste container.

Analyzing Evidence

1. What was the physical difference you saw in the test strips that allowed you to analyze different ozone concentrations? Rank your test sites according to the amount of ozone detected. Did you observe a large range of differences between ozone test sites?

2. Rank your test sites according to particulate concentrations. Can you identify the source of any of your particles based on their appearance? Did you observe any large differences in particle numbers or types between the sites?

Interpreting Evidence

1. What is the main source of ozone pollution at your test sites? How did your ozone concentrations change with respect to your independent variable?

2. What is the main source of particulate pollution at your test sites? How did the number and type of particulates change with respect to your independent variable?

3. Compare your results from each experiment to the hypothesis you wrote in *Making Predictions*. How accurately did you predict the results of each experiment? Write a revised hypothesis based on your results, and briefly explain how you would set up another experiment to test your new hypothesis.

Making Claims

What can you claim about sources of ozone or particulate pollution in each experiment? Support your statement with evidence from your experiment.

Reflecting on the Investigation

1. What factors other than pollution may have affected your results?

2. Did you see any unusual results in either experiment? If so, try to explain them.

3. Explain the difference between a chemical change and physical change. Which of these tests used a chemical change to detect the presence of a pollutant?

4. Smog City 2 (www.smogcity2.org) is an interactive air pollution simulator that allows you to see how variables such as weather, traffic, and land use affect ozone or particulate pollution. Go to the website and adjust the variables in the game to try to minimize the air pollution. Which variables are most important for minimizing ozone? Particulates?

5. If you are in the United States, go the AirNow website (www.airnow.gov) and look up the ozone and particulate pollution for your zip code. Record the AQI for particulate matter and ozone, and describe the health category into which the values fit. How does your area compare to other regions in the United States? *Note:* Many other countries have websites that report air quality, so if you are not in the United States see what is available for your country and find air pollution values for your community.

6. What changes could you make in your life to help improve air quality in your community?

Investigation adapted from *Chemistry in the Community*, 6th edition; W.H. Freeman, 2011.

Graphing the Mass of Air and the Temperature of Water

Asking Questions

- Air has a mass because it is composed of matter—elements and compounds. What are the compounds that you'll be weighing today?
- What is the relationship between the mass and pressure of air in a closed system?
- Do you expect the graphs you'll construct to have linear or curved regressions? Why?
- What are some advantages of using graphs to represent data?

Preparing to Investigate

In this experiment, you will collect data about air and water that lends itself to presentation in graphical form. You will then learn how to construct graphs in ways to make the visual presentations most effective.

Two short laboratory exercises are included. Your instructor will specify whether you are to do one or both. The first exercise studies the relationship between the pressure and mass of air in a closed container. The second investigates the relationship between time and temperature as a hot liquid cools.

Graphs present information pictorially, allowing you to see trends and relationships between the data you collect. The *Laboratory Methods* section describes in detail how to prepare graphs and analyze your data using **interpolation** and **extrapolation**. For this investigation you will do a regression analysis of your data. Carefully read the section on graphing in the *Laboratory Methods* section before doing the investigation.

Making Predictions

- After reading *Gathering Evidence*, for each exercise, sketch a graph that illustrates the relationship between the data sets. Identify whether the expected regression is linear or curved and explain why you predict that relationship.

- Prepare a data sheet that starts with the predicted graphs and includes space for observations and data tables for the assigned exercises.

Gathering Evidence

Overview of the Investigation

Part I. Weighing Air

1. Inflate a 2-liter bottle to about 40 pounds per square inch with a bicycle pump.

2. Weigh the bottle and record its mass. Record the pressure of the bottle.

3. Release some of the air and repeat Step 2 several times until all of the added air pressure has been released.

Part II. Cooling Water

1. Heat some water in a beaker to about 80 ºC.

2. Allow the water to cool and record its temperature at regular time intervals.

 STOP! Safety glasses must be worn *at all times* while doing chemistry investigations.

Part I. Weighing Air

1. For this investigation, you will use a laboratory balance for measuring the mass of an object. How to use a balance is describe in the *Laboratory Methods* section, and your instructor will explain the use of the particular balances in your laboratory.

2. Obtain a clean, dry plastic 2-liter bottle with a screw cap that has been fitted with a tire valve. Put a thermometer into the bottle and tightly screw on the cap.

3. Record the temperature of the air inside the bottle. Each time you change a variable, record the temperature from the inner thermometer. Also record your observations including whether the bottle feels warm or cool to the touch.

4. Become familiar with using the tire gauge to measure the air pressure. Measuring pressure is described in *Laboratory Methods*, and your instructor can assist you with your apparatus. Make sure you can push the gage onto the valve squarely without letting out much air and read the pressure units on the protruding stem. The pressure scale is often marked in pounds-per-square-inch (psi). Atmospheric pressure is 14.7 psi, and the tire gage measures pressures above (not under) atmospheric pressure.

5. Attach the hose from a hand tire pump to the valve. Place the bottle in a protective enclosure (such as a wastebasket). Pump air into the bottle until the pressure on the tire gage reads about 40 psi.

6. Measure and record the mass of the bottle. Wait 5 minutes and measure the mass again.

7. If the mass of the bottle (plus air) hasn't changed, you can start to collect data. If the mass of the bottle decreased by more than 0.05 g, tighten the cap and repeat Steps 5 and 6.

8. Carefully measure the air pressure in the bottle using the tire gage and record this in a data table. Weigh the bottle on a balance and record the mass (to the nearest 0.01 gram) and temperature.

9. Let a small amount of air out of the bottle. Measure and record the pressure and temperature again and then measure and record the mass.

10. Repeat step 9 until you have at least five sets of pressure, temperature, and mass data for points between your highest pressure and 10 psi.

11. Let all of the pressurized air out of the bottle, and measure and record the mass and the temperature. Remember that the bottle is not empty but now contains air, and should have a pressure of around 14.7 psi.

12. **OPTIONAL:** These measurements can be done very rapidly. If time permits and you are interested, you might like to collect a second set of data. Your data may be more accurate because measurement technique often improves with repeated investigations.

Part II. Cooling Water

 STOP! This part of the investigation involves an open flame. No flammable chemicals should be in the vicinity. Long hair must be tied back, and loose sleeves on clothing must be rolled up.

1. Use a thermometer or temperature probe to measure the air temperature in the room and record it.

2. Measure 50 mL of water using a graduated cylinder, pour it into a 100-mL beaker, and place the beaker on a ring stand over a Bunsen burner.

3. Light the Bunsen burner, heat the water to about 80 °C, and then turn off the Bunsen burner.

4. Allow the water to cool to 75 °C without stirring or removing the beaker from the ring stand.

5. When the water has cooled to 75 °C, start recording the temperature at intervals of 2 minutes. Use a wall clock, wrist watch, or stop watch to keep track of the time and record the temperatures to the nearest 0.5 °C.

6. Continue making and recording measurements until the water has cooled to about 35 °C.

Analyzing Evidence

A. Graphing Conventions

1. Graphs should be neat, legible, and well-organized.

2. The title should be descriptive and tell the reader what has been plotted, such as "Volume of plastic as a function of its mass." Titles such as "M vs. V" do not give enough information and should not be used.

3. The horizontal axis (*x*-axis) and the vertical axis (*y*-axis) should be clearly labeled to show what is plotted (*e.g.*, volume) and the units (*e.g.*, mL).

4. The *scale* of the graph should be chosen so that the graph fills as much of the paper as practical. In general, the scales on the *x*-axis and the *y*-axis do not need to start at 0. Figures 0.13 and 0.14 in the *Laboratory Methods* section show examples of correctly scaled graphs. The data points fill the graph space quite effectively. Figure 3.1 shows three examples of incorrectly scaled graphs.

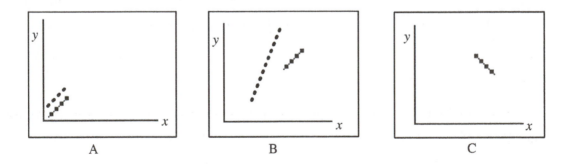

Figure 3.1. Examples of graphs with incorrect scales.

Graph A: The increments on the *x*- and *y*-axes should be made smaller to spread the graph out over the page.

Graph B: The starting point on the *x*-axis should be changed and the increments on the *x*-axis should be made smaller so the graph takes up most of the page.

Graph C: The starting point on both the *x* and *y* axes should be changed and the increments on both axes should be made smaller so the graph takes up most of the page.

5. If the data appear to represent a linear (straight-line) relationship, use a ruler to draw a single straight line that best represents the average relationship. If the data seem to follow a curve rather than a straight line, then draw the best *smooth* curve through them. It is unlikely that all of the points will fit exactly on one smooth line (either straight or curved); therefore, some judgment must be exercised in deciding on the "best" fit. You are trying to identify a *trend* in the data, so it is not effective to draw zig-zag lines that connect the data points.

6. When some data are plotted, it becomes apparent that certain points seem "out of line" with all others. Scientists often call these "outliers." In this case, it is likely that an error was made, either in the original measurement or in writing it down. (It is also possible that an error was made in placing the point on the graph—this is the first thing to check.) If a point really seems to be out of line with the others, it is appropriate to use your judgment and exclude it when determining the best line.

B. Graphing the Data

Part I. Weighing Air

1. Identify the range of recorded masses from lowest to highest. Select a *convenient* set of values that will include this range of masses. Use these to label the *y*-axis on the graph. You may want to select values with subdivisions of 10 since the masses are in decimal units.

2. Similarly, identify the range of recorded pressures. These data should be from 0 to about 40 psi. Choose a convenient way to divide the *x*-axis and label the axis.

3. Carefully plot the investigation points using a pencil. Make small dots, and then draw small circles around them so that they show up clearly.

4. Examine the data points carefully and decide whether a curved or straight line is most appropriate to represent the trend in your data. If the points are somewhat scattered, there should be equal numbers of them on either side of the line.

5. If specified by your instructor and if the line is straight, calculate the slope of the line. To do this accurately, DO **NOT** use particular data points. Instead, select two places on the line near the opposite ends of the line. Carefully read off the *x*- and *y*-values for each, and then use the equation given in the *Lab Methods* section to calculate the slope.

Part II. Cooling Water

1. Identify the range of recorded temperatures from highest to lowest. Select a *convenient* set of values that will include this range of temperatures. Use these to label the *y*-axis on the graph.

2. Select a convenient set of values for the *x*-axis, starting at zero, to cover the number of minutes over which you made measurements. Label this axis.

3. Carefully plot the investigation points using a pencil. Make small dots, and then draw small circles around them so that they show up clearly.

4. Examine the data carefully and sketch a trendline, as described in Step 5 in Part I.

Interpreting Evidence

1. An interesting further use of the data set from *Weighing Air* is to extrapolate the line to −14.7 psi. (If a computer-graphing program is available, this becomes easy to do simply by changing the scale on the horizontal axis.) What do you think you will see?

2. Suppose you could pump some air out of the bottle rather than pumping it in. Predict how the mass would change. Is there a limit to the change?

3. All of your mass measurements were actually the combined mass of the bottle itself plus the air it contained. How could you find out the mass of the air alone? *Hint:* What is the significance of the weight of the bottle at −14.7 psi?

4. Suggest a possible explanation as to why the bottle warmed and cooled when it did.

Making Claims

What can you claim about the relationship between the mass and pressure of air in a closed system? What can you claim about the relationship between temperature and time for water as it cools? Use the data and the graph to support your claims.

Reflecting on the Investigation

1. Write a paragraph describing the advantages of using a graph to represent data rather than just using a data table. Include examples encountered in this investigation's introduction and results.

2. Figures 4.19 and 4.21 in *Chemistry in Context* shows two plots of well-known data describing the increase in atmospheric carbon dioxide with respect to time. Study these figures and answer the following questions.

   a. What two quantities are plotted against each other in each graph? What are the units?

   b. The data from Mauna Loa is included in the plot spanning thousands of years. Why is it useful to have this data plotted separately, as in the inset graph in Figure 4.19?

   c. People continue to argue about the validity of extrapolating this data to predict future atmospheric carbon dioxide levels. In what ways might an extrapolation be valid, and what concerns may be raised by trying to extrapolate this data?

3. Choose another graph from your textbook, write down the figure number, and answer the following questions.

   a. What quantities are plotted against each other in the graph? What are the units?

   b. Describe what this graph tells you about the relationship between these quantities (*e.g.*, does one quantity increase or decrease as the other increases?). Does the graph indicate correlation, causation, or no relationship between these quantities? (Look up *correlation* and *causation* if you don't know what these terms mean.)

NOTES

Protection from Ultraviolet Light

Asking Questions

- Compare the energy of ultraviolet (UV) light photons to that of infrared and visible light photons. Why is UV light potentially more dangerous than visible or infrared light?
- What are the two ways that materials can protect us from UV light?
- What information would be important to know when designing a material to protect us from UV light?

Preparing to Investigate

Over the past few decades, we have become increasingly aware of the need to protect ourselves from exposure to UV light. As described in Chapter 3 of *Chemistry in Context*, UV light has sufficient energy to cause changes in DNA that can lead to skin cancers, and depletion of the ozone layer has allowed more UV light to reach the surface of Earth. A variety of materials—clothing, sunscreens, plastic coatings, and so on—provide protection from UV light, typically by either absorbing or reflecting it. With your laboratory team, you will design and conduct an experiment to test the utility of different materials to block UV light.

Making Predictions

Rank your proposed materials in order of their predicted ability to protect UV-sensitive beads from UV light, and explain why you chose this order.

Gathering Evidence

Designing the Investigation

In this investigation, you will use beads that change color when exposed to UV light. When the beads encounter more UV light, they change faster and the color becomes deeper. Choose at least three different materials and develop a procedure for testing them using the beads. Determine a quantitative method for comparing how quickly the beads change color and/or how dark the colors become.

You will collect your data outdoors. The beads are quite sensitive, though, so working in a shady spot usually gives better results. As you conduct your investigation, remember to protect the beads from all sides so that you are only measuring the exposure through the material. In between trials, put the beads in a dark place, such as your pocket; they will quickly change back to their original color.

Remember that good investigation design requires including control variables and multiple trials. You will want to determine how the beads respond when left totally unprotected. As you collect data, change only one variable at a time. Repeating measurements will allow you to report average data instead of single data points.

Optional extension: If you have access to a UV-vis spectrometer, obtain a spectrum of some of your tested materials, so you can see the absorbance of your materials at different wavelengths.

Analyzing and Interpreting Evidence

Keep a careful written record of what you do, and note any procedural changes from your original experimental plan, as well as why you made the changes. Record your data in appropriate tables, including both measured values and calculated averages. Identify which of the trials contained experimental controls and explain why they were important to include. Explain why you did multiple trials, and identify any results that would have been different if multiple trials were not conducted. Finally, rank your materials by their ability to protect against UV light, and justify your ranking based on your experimental evidence. Compare your results to your hypothesis, explaining any differences.

Making Claims

What can you claim about the ability of the materials you tested to protect from UV light? Can you draw any conclusions about the interactions between the different materials and UV light?

Reflecting on the Investigation

1. Describe one specific change that you would make in your procedure if you were doing the investigation again. Explain how this change could lead to better or more interesting results.
2. Do your materials block UV light by absorbing it or reflecting it? How do you know?
3. Most sunscreen lotions and sunglasses claim to protect against UVA and UVB light, but they don't mention UVC light (Table 4.1). Why? Is it less dangerous than the other types of UV light? What happens to the UVC light?

Table 4.1. Types of UV light, classified by wavelength.

UVA	320-400 nm
UVB	280-300 nm
UVC	100-280 nm

4. If you obtained UV-vis spectra of your materials, do they absorb light selectively at different wavelengths, or the same at all wavelengths? How does this indicate the blocking method?

NOTES

The Chemistry of Sunscreens

Asking Questions

- Compare the energy of ultraviolet (UV) light photons to that of infrared and visible light photons. Why is UV light potentially more dangerous than visible or infrared light?
- How do different compounds in sunscreens protect us from UV light?
- Are all sunscreen compounds equally effective at blocking UV light?
- Why do commercial sunscreens usually contain multiple active ingredients?
- What is meant by the claim that sunscreens provide either UVA or UVB protection, or both?
- What types of information would you need to be able to design an effective sunscreen?

Preparing to Investigate

In this investigation, you will make sunscreens by mixing a variety of compounds that are used in commercial sunscreens with lotion. You will then test the sunscreens' abilities to absorb ultraviolet (UV) light. You will also measure the absorbance of UV light by these compounds using a UV-visible spectrophotometer (see the section in the *Laboratory Methods* section of this lab manual).

As described in Chapter 3 of *Chemistry in Context*, light in the ultraviolet range of the electromagnetic spectrum has shorter wavelengths and more energy than infrared or visible light. UV light has sufficient energy to cause changes in DNA that can lead to skin cancers, and depletion of the ozone layer has allowed more UV light to reach the surface of Earth. Awareness of the dangerous components of sunlight has led to the development of materials that can reduce our exposure while allowing us to enjoy time outside.

One strategy to protect ourselves from UV light is to use sunscreen. A variety of commercial lotions are available with sun protection factors (SPFs) ranging from 2 to 50. Many of these sunscreens contain organic molecules, such as oxybenzone or octyl methoxycinnamate, which absorb light specifically in the UV region of the spectrum. Some sunscreens contain small particles (nanoparticle) of titanium dioxide or zinc oxide, which are designed to scatter the light so the skin can't absorb the rays. The lenses in sunglasses often absorb UVA, UVB, and even UVC light (Table 5.1).

Table 5.1. Types of UV light, organized by wavelength.

UVA	320-400 nm
UVB	280-300 nm
UVC	100-280 nm

Making Predictions

- Look at the labels of several commercial sunscreens. What compounds are included on the list of active ingredients? What is the range of concentrations (% w/w) for the compounds?

- UV-visible data show the wavelengths where a certain compound absorbs light. Read the Spectroscopy section in the *Laboratory Methods* section of this lab manual, and then answer the following question:
    - What do you expect the UV-visible spectra to look like for compounds used in sunscreen? Explain your prediction.

- After reading *Gathering Evidence*, prepare a data sheet to record your weight percentage data, observations of UV absorption on test squares, and data from UV-visible spectra.

Gathering Evidence

Overview of the Investigation
1. Prepare lotions that contain 5% w/w of test compounds.
2. Make test cards by drawing squares on index cards with a fluorescent yellow highlighter.
3. Apply lotions to the squares and use a UV lamp to test their ability to absorb UV light.
4. Measure how the lotion samples, the base lotion, and a commercial sunscreen absorb UV light by collecting spectra using a UV-visible spectrophotometer.

Part I. Making Lotion Samples Containing 5% w/w Test Compounds
1. Your instructor will notify you of the active ingredients from sunscreen that are available for this investigation. Select three organic molecules to test. Record the name and formula of each compound. Also, record the formula and size of any nanoparticles that you will be testing.
2. Label a small weighing boat for each sample.

3. For the first sample, place the weighing boat on the balance, tare the balance, add a small amount of lotion to the weighing boat, and record the mass of the lotion on your data table.

4. Calculate the amount of compound to add to make a 5% w/w composition by multiplying the mass of the lotion by 0.05263. Record the mass of the compound on your data table.

5. Weigh the calculated amount of compound on weighing paper and add it to the lotion. Mix well using a cotton swab.

6. Repeat Steps 3-5 for the rest of the samples.

Part II. Testing UV Absorption of Sunscreens

1. Make a test card for each compound by drawing three squares in line across an index card using a fluorescent yellow highlighter. Label the first square with <u>Lotion</u>, the second square with the name of the compound, and the third square with <u>Sunscreen</u>.

2. Use the cotton swab to apply some of a lotion sample to the middle square of the appropriately labeled card. The coating should be applied as if you were putting sunscreen on your skin (not too thick and not too thin).

3. Repeat Step 2 for all lotion samples.

4. Put a small amount of lotion and commercial sunscreen in separate small weighing boats. Apply lotion (negative control) to the first square and commercial sunscreen (positive control) to the third square on each card.

5. Record your observations on your data table.

6. Place each card under a UV lamp and record your observations on your data table. Be sure to compare the absorption of each lotion sample to both controls and to other samples.

7. If your UV lamp has multiple bulbs that produce light of different wavelengths, use each source to collect data.

8. Predict which compounds will absorb light in the ultraviolet region and, if possible, which will absorb in UVA light and which will absorb UVB light.

Part III. Measuring UV Absorbance of Test Compounds

1. Tare a labeled scintillation vial on a balance and add approximately 10 mg of one of the test compounds. Record the exact mass on your data table.

2. Add 10 mL of 2-propanol. Put the cap on the vial and shake it to dissolve the compound.

3. Repeat Steps 1 and 2 for each test compound.

4. Your group will be given a quartz cell to use when measuring UV absorbance (while glass blocks UV light, quartz is transparent throughout the UV-visible region).

CAUTION! These cells are expensive. Be careful with them, and only handle them by their frosted sides to avoid leaving finger prints behind.

5. Fill the cell about three-quarters of the way with 2-propanol.

6. Put the cell in the cell holder of the spectrophotometer, and collect a blank spectrum of the 2-propanol. This spectrum will be saved and subtracted from the sample spectrum.

7. Remove the cell and add 1 drop of the first test solution. Use the pipet to mix the solution by gently drawing solution into the pipet and dispensing it back into the cell several times.

8. Place the cell back in the holder and collect a sample spectrum of the test solution.

9. If the highest peak in the absorbance spectrum is between 0.7 and 2 absorbance units, save the spectrum and skip to Step 12.

10. If the highest peak in the absorbance spectrum is below 0.7 absorbance units, increase the concentration of the test compound in solution by repeating Steps 7–10 until you are able to collect data within the range of 0.7 to 2 absorbance units. Keep track of how many cycles you needed for each compound to get to a measurable concentration.

11. If the highest peak in the absorbance spectrum is above 2 absorbance units, remove the cell from the holder and dilute the solution in the cell. Use the pipet to remove solution until the cell is about one-quarter full. The excess solution can be added to the scintillation vial for that solution. Add 2-propanol to refill the cell to about the three-quarter level and mix with the pipet. Repeat Steps 8–11 until you are able to collect data within the range of 0.7 to 2 absorbance units. Keep track of how many cycles you needed for each compound to get to a measurable concentration.

12. Record in your data table both the wavelength range where absorbance occurs and the specific wavelength that shows the highest absorbance. Rinse the cell thoroughly with 2-propanol. Collect all rinses in a waste beaker.

13. Start at Step 5 for each of the remaining test solutions.

Part IV. Measuring UV Absorbance of Control Lotions

1. Tare a 13x100 mm test tube and use a pipet to transfer 2 drops of the commercial sunscreen to the bottom of the test tube. Record the mass of the sunscreen.

2. Add 5 mL of 2-propanol to the test tube and use a stirring rod to mix the sunscreen and solvent. The active ingredients should be soluble in 2-propanol, but other parts may not dissolve. The solution should be either finely dispersed or fully dissolved.

3. Repeat Steps 1 and 2 using the lotion that contains no added compounds.

4. Centrifuge the test tubes for 3–4 minutes. This should force anything that has not dissolved into the bottom of the tube.

5. Fill the quartz cell to about the three-quarter level with 2-propanol.

6. Put the cell in the cell holder of the spectrophotometer, and collect a blank spectrum of the 2-propanol. This spectrum will be saved and subtracted from the sample spectrum.

7. Remove the cell and empty it into the waste beaker. Without disturbing the solid at the bottom of the test tube containing the lotion, pipet enough of the solution to fill the cell to the three-quarter level.

8. Place the cell back in the holder and collect a sample spectrum of the test solution.

9. If the highest peak in the absorbance spectrum is below 2 absorbance units, save the spectrum and skip to Step 11.

10. If the highest peak in the absorbance spectrum is above 2 absorbance units, remove the cell from the holder and dilute the solution in the cell. Use the pipet to remove some of the solution. The excess solution can be put back in the test tube by dripping it down the side. Add 2-propanol to refill the cell to about the three-quarter level and mix with the pipette. Repeat Steps 8–10 until you are able to collect data within the range of 0.7 to 2 absorbance units. Keep track of how many cycles you needed for each compound to get to a measurable concentration.

11. Record ion your data table both the wavelength range where absorbance occurs and the specific wavelength that shows the highest absorbance. Rinse the cell thoroughly with 2-propanol. Collect all rinses in a waste beaker.

Clean up

As directed by your instructor, discard or recycle the contents of your scintillation vials, test tubes, and waste beaker and put weighing boats and test cards into the solid waste container or trash. Rinse your quartz cell and return it to your instructor.

Analyzing Evidence

1. What did you observe when the squares for the lotion samples were exposed to visible light? What did you observe when they were under UV light?

2. Identify the wavelength that had the highest absorbance for each test compound. Order the compounds from the one with the shortest maximum absorbance wavelength to the one with the longest maximum absorbance wavelength.

3. Which of the test solutions for UV-visible spectroscopy, if any, needed to be diluted before being measured? Which of the test solutions for UV-visible spectroscopy, if any, needed to be more concentrated before being measured?

Interpreting Evidence

1. You made a prediction as to what the UV-visible spectra for compounds used in sunscreens would look like. Was your prediction correct? If yes, what basis did you use for the prediction? If no, what error did you make?

2. Which of the test compounds absorb UV light? Use evidence from your results to support your answer.

3. Divide the test compounds into four groups – those that absorb UVA light, those that absorb UVB light, those that absorb both UVA and UVB light, and those that are not UV absorbers.

4. Use the answer to Question 3 in Analyzing Evidence to rank the compounds you tested from weakest absorber to strongest absorber. Explain your ranking.

Making Claims

What can you claim about the active ingredients of sunscreen?

Reflecting on the Investigation

1. You observed in this investigation that compounds in sunscreen absorb different wavelengths of UV light and demonstrate different amounts of absorption. What data would you use to design an effective sunscreen? Which of the compounds you tested would you include? What relative concentrations would you use? Use your data from the investigation to support your decisions.

2. You included lotion as a negative control and a commercial sunscreen as a positive control in your investigation. What did the results from these products show? Why was it important to include each of these?

Investigation adapted from

1) Moeur, H. P.; Poon, T.; Zanella, A. An Introduction to UV-Vis Spectroscopy Using Sunscreens. *J. Chem. Ed.* **2006**, *83*, 769.

2) Guedens, W. J.; Reynders, M.; Van den Rul, H.; Hardy, A.; Van Bael, M. K. ZnO-Based Sunscreen: The Perfect Example To Introduce Nanoparticles in an Undergraduate or High School Chemistry Lab. *J. Chem. Ed.* **2014**, *91*, 759-263.

Color and Light

Asking Questions

- Why are some objects colored? How are they interacting with light?
- What is spectroscopy?
- What is the relationship between the color of an object and the color(s) of the light that it absorbs?

Preparing to Investigate

What is a rainbow? When white light, which is a mixture of electromagnetic radiation of different wavelengths, passes through water in the atmosphere or through a prism, it separates into the different component waves. These wavelengths of light are detected by human eyes as color, and therefore are part of what is called the visible region of the **electromagnetic spectrum** (Chapter 3 in *Chemistry in Context*). Each color has a unique position in the electromagnetic spectrum that is identified by its **wavelength**. In the visible region, the wavelengths are in the range of billionths of a meter and are expressed in **nanometers** (1 nm = 1×10^{-9} meter).

In this investigation, you will observe the separation of visible light and interactions of light with colored solutions. After determining the colors of different wavelengths of light in the visible region, you will measure the wavelengths of visible light that are transmitted or absorbed by colored solutions. You will construct a **spectrum** for each substance that shows how much light of different wavelengths is absorbed or transmitted.

You will use a **spectrophotometer** for this investigation. The components and use of a spectrophotometer are described in detail in the *Laboratory Methods* section, and you should read that section before proceeding to the investigation. The operation of each spectrophotometer is different, and so you will be given detailed directions from your instructor for the operation of your specific instrument. Spectrophotometers are sensitive and expensive instruments, so it is important to follow the instructions carefully to avoid damaging the equipment. Your instructor will let you know whether to measure the % transmittance (%T) or absorbance (A) of the light.

Making Predictions

- Which of the solutions, the red, blue, or green, do you expect to absorb light at the shortest wavelength? Longest wavelength? Explain why you predict that relationship.

- After reading *Gathering Evidence*, prepare a data sheet that includes data tables for the spectroscopic measurements and space for observations, predictions, and graphs.

Gathering Evidence

Overview of the Investigation

1. Prepare a piece of chalk to reflect light from the spectrophotometer source.
2. Determine the color of light of selected wavelengths from 400 to 700 nm.
3. Measure the transmittance (or absorbance) spectrum for a red solution and a blue solution.
4. Predict and then measure the spectrum for a green solution.
5. Plot a graph of transmittance (or absorbance) vs. wavelength for each colored solution.
6. Optional: Use this procedure to investigate a colored substance of your choosing.

Part I. Assigning Wavelength to Color

1. Turn on the spectrophotometer and let it warm up. Your instructor will let you know whether to choose the %T or A mode, and show you how to adjust the wavelength. Refer to *Laboratory Methods* for more details.

2. Take a half-inch long piece of chalk and rub it on a blackboard or another durable surface until one end is worn down to a forty-five-degree angle (Figure 6.1). Place the chalk in the test tube and insert it into the spectrophotometer so that the slanted side is pointing toward the light source.

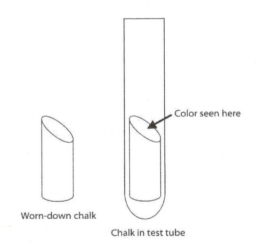

3. Adjust the wavelength to 500 nm.

4. Look down the tube to the slanted piece of chalk. You should see a colored band of light. If you don't, raise, lower, or rotate the test tube slightly until you do. If you still cannot see the colored band, consult your instructor for further assistance.

Figure 6.1. Chalk prepared for the spectrophotometer.

5. Slowly adjust the wavelength in both directions and observe that the color of the light band changes.

6. If you change the wavelength far enough in either direction, you will not see any color because the wavelength will be beyond the region that your eye can detect. By careful observation, find out the limits for *your* eyes and record these limits.

7. Finally, find out which wavelengths correspond to which colors. To do this, return the wavelength to 400 nm. Look down the tube and record in your data table the color you see. Change the wavelength to 440 nm and again record the color you see. Repeat this at 40 nm intervals from 400 to 720 nm.

Part II. Collecting Transmission or Absorbance Data for Red and Blue Solutions

1. Obtain three test tubes or other containers, often called cuvettes, that fit the spectrophotometer you will be using. Fill one of them about half full with a red dye solution. Fill another container half way with a blue dye solution. Fill the last container half full with pure water. The water will be the blank, the other two the sample tubes.

2. Put the blank in the instrument, and if necessary, adjust the meter on the spectrophotometer to read 100% T or 0 A, depending on which mode you are using.

3. Remove the blank, insert the sample tube with the red solution and close the cover. *Without changing any knobs or pressing any buttons*, carefully observe the meter reading and record this in the data table as % T (or A) for the red solution at 400 nm.

4. Place the other sample tube containing the blue solution in the instrument, close the cover, and read the % T (or A) and record your measurement.

5. Change the wavelength dial to 420 nm. Repeat Steps 3 and 4 by first inserting the blank, closing the cover, and zeroing the instrument. Then insert the samples, observe the meter reading, and record your measurement. Repeat with the blue solution.

6. Proceed in the same fashion, at 20-nm intervals, to 700 nm. Although it may sound tedious, data collection goes quite rapidly when two students work together. Remember that *each* time you change the wavelength, you must insert the distilled water blank and adjust to 100% T or 0 A. Then collect the data for both the red and blue solutions.

7. Finally, make a *prediction* of what you think the spectrum will look like for a green solution, using the data from the red and blue solutions. Show it to your instructor before proceeding.

8. Rinse and fill the sample tube with a green solution and proceed to record the transmittance or absorbance data for this solution in the same manner as for the other solutions. Make sure to use the blank for each measurement.

Part III. Spectrum of an Indicator Solution

1. Rinse a sample test tube and fill it halfway with bromothymol blue indicator solution. A small amount of sodium hydroxide has been added to this solution in advance to be sure the solution is alkaline. Rinse and half-fill another test tube or cuvette with the same indicator solution. Add drops of hydrochloric acid solution to make the solution acidic. It should change the indicator color to yellow when it's acidic.

2. Follow the same procedure that you did in Part II. At each wavelength chosen, first insert the blank test tube and set to 100% T or 0 A. Then measure the colored solutions from 400 nm to 700 nm. Be sure to measure both test tubes (acidic and basic) before proceeding to the next wavelength.

3. Find a way to make an intermediate form of the indicator solution that is green. You will have to experiment by adding small amounts of very dilute acid (hydrochloric acid) and/or very dilute base (sodium hydroxide) to achieve a green color. Predict what the spectrum will look like, then measure the spectrum of this green solution and compare it with the spectra of the blue and yellow forms.

Part IV. Investigating Other Colored Solutions

There are many colored substances that you could use in a spectrophotometric study. Some common examples include beverages, such as sports drinks, soft drinks, or juices, as well as water-color paints, felt-tip pen inks, food color dyes, or hair dyes. You could even study small pieces of colored glass if they can fit into the sample chamber. Whatever you choose must be transparent (you can see through it) and fairly bright, and distinctly colored.

You may have to experiment to find a suitable dilution of your substance so that the lowest transmittance does not drop below 5% T (or absorbance above 1.5). Then use the skills you have learned previously to collect data and plot a spectrum of the substance.

Part V. Collecting Ultraviolet or Infrared Spectra

Your instructor may demonstrate or allow you to use a spectrophotometer that measures ultraviolet or infrared spectra. You may be able to obtain an ultraviolet spectrum of a sunscreen you analyzed in Investigation 5. You also may be able to obtain an infrared spectrum of a greenhouse gas, such as carbon dioxide.

Analyzing Evidence

1. Plot the data for each colored solution or substance. It is easy to do this by hand on graph paper but you can also use a computer program. The wavelength should be on the horizontal

axis (*x*-axis), ranging from 400 nm to 700 nm. Percent transmittance or absorbance should be on the vertical axis (*y*-axis); % T should range from 0 at the bottom to 100 at the top or absorbance should range from 0 at the bottom to 1.0 (or 2.0) at the top.

2. First plot the red solution. *Use a pencil so that corrections can be made.* Carefully plot the data, making a small point with a circle around it for each measurement. When you are finished, draw a *smooth* curved line through the points. The line may not touch every point.

3. Repeat this procedure with other colored substances. You can plot more than one spectrum on the same graph, but, if more than one spectrum is on as single graph, label each spectrum clearly or use a different color pencil for each line.

Interpreting Evidence

1. How do the spectra change when measuring red, green, and blue solutions?
2. At what wavelengths does each solution have its greatest transmittance (lowest absorbance)? What color(s) of light correspond to these wavelengths?
3. At what wavelengths does each solution have its lowest transmittance (highest absorbance)? In these regions, the dye molecules are absorbing the light. What color(s) of light is the red/blue/green dye solution absorbing?
4. How does the spectrum of the green dye solution differ from that of the green indicator?

Making Claims

What can you claim about the color of an object and the color(s) of the light that it absorbs?

Reflecting on the Investigation

1. Humans can typically see light in the range of 400 to 700 nm. However, some animals can see wavelengths outside of this range. For instance, bees and some spiders can see ultraviolet light, while snakes and other reptiles have vision extending to the infrared wavelengths. What adaptive advantages might this extended vision offer these animals?
2. Why is the sky blue? And why does it appear red or orange at sunrise and sunset?
3. Our eyes see color by using light-sensitive cones. We have three types of cones that each see a different color—red, green and blue. If we have only these three types of cones, how can we see more than just three colors? People who are color-blind lack one or more type of cones. Explain how vision is affected for someone who lacks red cones.

4. Mixing colored light is different than mixing colored paint. For instance, if you mix red and yellow paint, you get orange, and if you add blue to the orange you get brown or black. However, your computer screen and television make colors by mixing light. Mixing red and green light makes yellow light, and adding blue light makes white. Investigate and write a paragraph about how and why mixing light is different than mixing paint.

NOTES

Testing Refrigerant Gases

Asking Questions

- How do refrigerant gases work?
- Which compounds were traditionally used as refrigerant gases before chlorofluorocarbons (CFCs)?
- What properties of CFCs make them ideal refrigerant gases?
- Why were CFCs outlawed in the 1990s?
- Which compounds have taken the place of CFCs as modern refrigerant gases?

Preparing to Investigate

Refrigerators and air conditioners take advantage of the physical properties of liquids and gases. Heat energy is given off when a vapor condenses to a liquid. Conversely, heat energy is absorbed when a liquid vaporizes. The cooling mechanism in a refrigerator or air conditioner is a closed system, which contains a compound that boils at a low temperature but can be converted easily back to a liquid under pressure. The liquid is allowed to vaporize in metal tubes inside the refrigerator. As it vaporizes, it takes heat from its surroundings and cools the inside of the refrigerator. The vapor produced is then pumped to metal tubes on the rear of the refrigerator, outside the cold compartment, and it is converted back to a liquid using pressure applied by a compressor. As the gas is condensed, it loses the heat it picked up inside the refrigerator and makes the coils on the back of the refrigerator feel warm. Through this process, heat is pumped out of the refrigerator and into the kitchen. In an air conditioner, the heat removed from the air in a room is pumped to the outside.

To efficiently carry out this refrigeration cycle, refrigerant gases must have certain physical properties: their boiling point must be well below 0 °C (so they do not condense back to a liquid inside the refrigerator or freezer compartment), they must be easily liquefied under pressure, and they should be non-flammable. Before CFCs, the most common gases that met these requirements were ammonia (NH_3), which boils at –33 °C, and sulfur dioxide (SO_2), which boils at –10 °C. Both can be liquefied at room temperature under high pressures. At room temperature (75 °F, 24 °C), ammonia becomes a liquid at a pressure about 10 times greater than atmospheric pressure, and sulfur dioxide will liquefy at about five times atmospheric pressure.

Today, it is hard to understand why the discovery of CFCs was hailed as one of the wonders of modern chemistry. These compounds are now viewed with great concern because of

molecules that you studied in this investigation, and rotate them using the computer program. What advantages do you see for viewing molecules this way as compared to the pictures in your textbook? What advantages and disadvantages do you see for these computer images as compared to the physical models you constructed in lab?

3. You will revisit many of these molecules as you go through the course. Answer the following questions about some of the molecules you studied in this investigation.

   a. Define the term "greenhouse gas" and explain how CO_2 functions as one. What other molecules that you studied today can act as greenhouse gases?

   b. What role does O_3 play in the atmosphere? What molecules are involved in its formation?

   c. Explain the problems with the use of CFCs as refrigerants. What class of molecules has replaced them?

   d. What physiological role does NO play?

Extracting Limonene with Liquid CO$_2$

Asking Questions

- What phase changes occur in CO$_2$ as temperature and pressure increase?
- Why is it important that a substance be soluble in the solvent used for extraction?
- What properties of supercritical and liquid CO$_2$ allow these solvents to be classified as environmentally friendly?

Preparing to Investigate

Essential oils are organic compounds extracted from natural sources such as fruits, herbs, and spices. They are used as flavorings and fragrances in a variety of products. Traditionally, essential oils are isolated through the process of steam distillation or by extraction with organic solvents. The process of **extraction** separates a compound or small number of compounds from a more complex mixture through differences in solubility. When a particular solvent is introduced, the soluble compounds dissolve in the solvent while the insoluble compounds do not. The compounds can then be separated from each other. In this investigation, you will extract a compound called *limonene* (Figure 9.1), a major component of the flavor and odor of oranges, from orange rind.

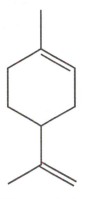

Figure 9.1. Structure of limonene.

Many traditional extraction methods have serious environmental drawbacks, including intensive energy-use and the production of large amounts of toxic waste. The past few decades have led to an increase into the investigation and use of supercritical CO$_2$ in place of organic solvents. This solvent is nonflammable, relatively nontoxic, and environmentally benign. Additionally, the properties of the solvent can be altered and used for a variety of applications by simply changing the temperature and pressure of the CO$_2$.

We are most familiar with CO$_2$ in the form of a gas, as it exists in the atmosphere or as we exhale it. Some of you may have seen solid CO$_2$, which is also known as dry ice. Unlike water ice, which melts to a liquid, dry ice **sublimes,** which means it goes directly from a solid to a gas. This is explained further in the phase diagrams below (Figure 9.2).

Phase diagrams describe the form a substance takes at different temperatures and pressures. Figure 9.2(b) shows the phase diagram for water. If you draw a line at atmospheric pressure (1 atm) from −100 °C to 400 °C, you'll notice that solid water becomes a liquid (at 0 °C) and then a gas (at 100 °C) as the temperature increases. If you were to reduce the pressure at 0 °C, you reach the **triple point**, where water exists in all three phases. At pressures below the triple point, liquid water cannot exist, and ice would sublime directly to water vapor. If you were to heat the water to 374 °C and raise the pressure above 218 atm, you would reach the **critical point**. Above this temperature and pressure you have a **supercritical fluid**, where the sample exhibits both gas and liquid characteristics. It will completely fill a container like a gas but will dissolve substances like a liquid does. Supercritical fluids occur naturally at underwater ocean vents and in the atmospheres of gas giant planets such as Jupiter and Saturn.

In the phase diagram for carbon dioxide (Figure 9.2(a)), you can see that the triple point occurs at 5.1 atm and −57 °C. This means that atmospheric pressure is less than the triple point pressure, so liquid carbon dioxide cannot exist at atmospheric pressure. This is why dry ice sublimes. You will also notice that the critical point for CO_2 occurs at a much lower temperature and pressure than that of water, −31 °C and 7.4 atm, which makes it quite easy to generate. Therefore, supercritical CO_2 has replaced more toxic solvents in applications such as dry cleaning and decaffeination of coffee.

To form liquid CO_2 as part of this investigation, we must raise the pressure of the carbon dioxide above the triple point pressure. We will do so by sealing dry ice inside a plastic tube, where the pressure from the subliming CO_2 will become high enough to generate the liquid form. You will use liquid CO_2 to extract limonene from orange rind. CO_2 is a nonpolar solvent, so it is appropriate to use it to dissolve nonpolar compounds, such as limonene. You will isolate the essential oil of orange, a mixture of compounds of which greater than 90% is limonene.

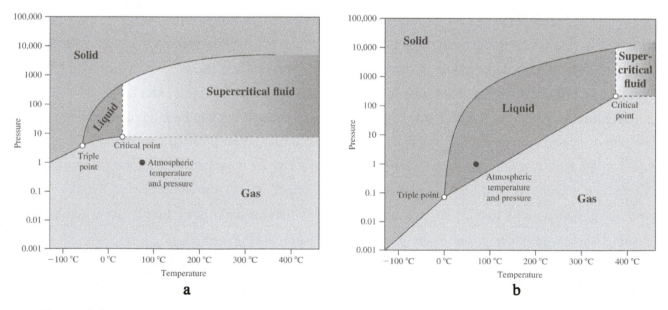

Figure 9.2. Phase diagrams for (a) carbon dioxide and (b) water.

Making Predictions

- Predict what percentage of the initial mass of orange rind will be recovered as oil.

- After reading *Gathering Evidence*, prepare a data sheet that includes space for your predicted percent recovery, the initial mass of the tube, the mass of the added orange peel, the mass of the tube with extracted limonene, the mass of the extracted limonene, and measured percent recovery. Leave room for your observations and calculations of the final mass and percent recovery.

Gathering Evidence

Overview of the Investigation

1. Prepare your centrifuge tube by making and inserting a coiled copper wire.

2. Grate orange peel and place it on top of the coil.

3. Fill the tube with dry ice, cap it, place in cylinder of warm water, and watch the extraction.

4. Record your observations and measure the amount of limonene produced.

Part I. Preparing the Sample

 STOP! Safety glasses must be worn *at all times* while in the chemistry laboratory. Do NOT use any glass containers during this investigation.

1. Using a balance, measure the mass of a clean, dry 15-mL plastic centrifuge tube. Record the mass on your data sheet.

2. Grate the colored part of the peel of half an orange using the smallest grating surface of a cheese grater. Weigh the peel using a balance and record the mass on your data sheet. It should be approximately 2.5 g.

3. Use a 20-cm piece of copper wire to make a trap for the solids by coiling the wire tightly at one end so that it fits in the tube at the

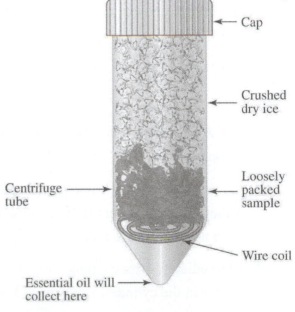

Figure 9.3. Extraction setup.

Labels: Cap, Crushed dry ice, Loosely packed sample, Centrifuge tube, Wire coil, Essential oil will collect here

top of the taper (Figure 9.3). This will prevent your solid from falling to the bottom of the centrifuge tube and interfering with recovery of the essential oil. The other end of the wire should extend upward but not past the top of the tube. The wire should be completely inside the tube so that the cap can be sealed.

4. Place your orange rind sample into the tube, making sure that none of it falls to the bottom of the tube. Do not pack it tightly.

Part II. Extracting the Oil

 CAUTION! Due to the safety issues involved with the rapid increase of pressure during this procedure, it is important to read and understand the *entire* extraction procedure (Steps 1–8) before beginning the experiment.

1. Fill a clear plastic cylinder about two-thirds full of warm tap water. Place the cylinder on the lab bench and move any items that should not get wet away from the cylinder. Splashing may occur if the top of your reaction tube comes off during the extraction.

 CAUTION! Do not heat the water in the cylinder or add hot water later in the procedure. Any sudden increase in temperature of the surrounding water when the tube is under pressure can cause the cap to blow off suddenly and violently.

2. Use a thermometer to measure the temperature of the water, and record it on your data sheet. Remove the thermometer from the cylinder before continuing.

3. Fill the centrifuge tube to the top with crushed dry ice. Tap the tube on the bench and continue to add more dry ice until the tube is full.

 CAUTION! Wear gloves when handling dry ice because direct contact can damage skin tissues.

4. Twist the cap on tightly until it stops turning. *Note:* If the cap never stops turning or otherwise appears not to fit the tube well, do not proceed. You will need to use a new cap or a new centrifuge tube.

5. Immediately after capping, drop the centrifuge tube, tapered end down (cap up), into the water in the cylinder. Pressure will begin to build in the tube.

CAUTION! Watch the extraction from the side, not the top! If the tube shatters or the cap shoots off any projectiles will be directed straight up. The plastic cylinder functions as a secondary container and protects you from possible injury.

6. After about 15 seconds, liquid CO_2 should be visible. The entire extraction will take approximately 3 minutes. If no liquid has appeared after 1 minute, there is not a sufficient seal on the tube. Carefully remove the tube from the cylinder, tighten the cap, and put the tube back in the cylinder. If repeated trials don't produce liquid, try replacing the cap and/or the tube. *Never remove the tube from the plastic cylinder when the CO_2 is liquid.*

7. After the liquid has evaporated and the gas is no longer escaping, remove the tube from the cylinder and open the cap.

CAUTION! Open the tube slowly and only after the gas has escaped. Do not point the cap at anyone, including yourself, as you are opening it.

8. Repeat the extraction once or twice more by adding dry ice to the tube as explained in Step 3 and the subsequent steps above. After two or three extractions, you should see a small amount of liquid at the bottom of the tube. This is the essential oil.

9. After your final extraction, carefully remove the coiled wire and the solid. If any solid remains in the tube, remove it with a spatula. Dry the outside of the tube with a paper towel. Measure and record the mass of the tube and the oil on your data sheet, along with any observations about the oil. Your instructor will demonstrate the proper way to observe the odor of the oil. Measure and record the temperature of the water in the cylinder.

Clean up

Discard or recycle the contents of your test tubes as directed by your instructor. Clean your thermometer with soap and water to remove all traces of the oil.

Analyzing Evidence

1. Calculate the mass of the extracted oil by subtracting the mass of the empty centrifuge tube from the mass of the tube with the oil.

Mass of essential oil = (mass of tube with oil) – (mass of empty tube)

2. Use the calculated mass of oil to determine how much oil is present in a particular mass of rind (percent recovery). Calculate the percent recovery of your essential oil by dividing the mass of the oil by the mass of the orange rind that you put in the tube.

$$\text{Percent recovery} = \frac{\text{mass of essential oil}}{\text{mass of orange rind used}} \times 100\%$$

Interpreting Evidence

1. Explain the differences between the *liquid* CO_2 that you generated in this investigation and *supercritical* CO_2 that is used as a commercial solvent.

2. Why was it necessary to add warm water to the cylinder that acted as a secondary reaction vessel? Consider that dry ice sublimes at –78°C at 1 atm. What is the minimum temperature at which liquid CO_2 can be generated?

3. How did the temperature of the water change during the extraction—did it increase or decrease? Explain your observations.

4. What was your percent recovery? Does this seem low or high for an extraction of essential oil? Why?

5. Identify several factors that could have affected your yield of oil and explain whether they would have made the yield higher or lower.

Making Claims

What can you claim about the use of carbon dioxide to extract essential oil from a natural product?

Reflecting on the Investigation

1. Limonene is a member of a class of molecules called *terpenes*, which are composed of one or more five-carbon *isoprene* units. Terpenes are important natural products, used in cleaning products and as organic pesticides. Some vitamins and other important biomolecules are also terpenes.

    a. How many isoprene units are present in limonene? *Hint:* Count the carbons.

b. Another common and commercially important terpene is menthol (right), a major component of peppermint oil. You may be familiar with this compound as a major ingredient of Vicks VapoRub™ and similar products. How many isoprene units are present in menthol?

c. Retinal is a light-sensitive compound that plays an important role in vision. The formula for this compound is $C_{20}H_{28}O$. How many isoprene units are found in this compound?

d. Our bodies make retinal from β-carotene, the orange compound found in carrots and other vegetables. The chemical formula for β-carotene is $C_{40}H_{56}$. How many isoprene units does it contain?

e. If you have covered Chapter 12 in *Chemistry in Context*, identify the functional groups present in limonene and menthol.

2. The introduction to this investigation states that carbon dioxide is a benign solvent, but you have probably read in *Chemistry in Context* that CO_2 is a greenhouse gas and can lead to global warming as well as contribute to ocean acidification. Normally, no new CO_2 is produced to make dry ice and supercritical CO_2. Rather, it is produced as a by-product of other processes. Search the web to find out where it comes from, and explain why it does not generally contribute additional CO_2 to the atmosphere.

3. Supercritical CO_2 has replaced chlorinated organic solvents such as methylene chloride and perchloroethylene (PERC) for applications such as dry cleaning and decaffeination of coffee. These solvents were initially chosen for their ability to dissolve a variety of substances and because they are not flammable, making them relatively safe to work with compared to hydrocarbon solvents. Search the web for these chlorinated solvents, identify some of their hazards, and explain why CO_2 may be a better option. You may wish to search for the Safety Data Sheet (SDS), which records important information about chemical safety.

4. The use of liquid and supercritical carbon dioxide is a large area of research for green chemists. Read the introduction to green chemistry in this book and take a look at the key ideas of green chemistry on p. viii. Identify which key ideas apply to the use of supercritical CO_2 as an industrial solvent.

Investigation adapted from McKenzie, L.C.; Thompson, J.E.; Sullivan, R.; Hutchison, J.E. Green chemical processing in the teaching laboratory: a convenient liquid CO_2 extraction of natural products. *Green Chemistry* **2004**, *6*, 355-358.

Measuring Molecular and Molar Mass

Asking Questions

- How can we measure the mass of a molecule?
- Do all gases have the same molar mass?
- What are some the relationships between mass and volume of gases?

Preparing to Investigate

In Chapter 4 in *Chemistry in Context*, you learned about **molar mass**—the mass (in grams) that contains 6.02×10^{23} (Avogadro's number) atoms or molecules of an element or compound. This investigation provides an opportunity to measure the molar masses of several gases and compare your investigation results with the accepted values for those gases.

Equal volumes of gases at the same temperature and pressure contain equal numbers of particles (atoms or molecules). This means that comparing the masses of equal volumes of two gases is the same as comparing the masses of equal numbers of particles of the substances. The *ratio* of the masses of an equal number of molecules therefore will be the same as the ratio of the masses of the *individual particles* and the ratio of the *molar masses* of the substances involved. This can be written as

$$\frac{\text{mass of } n \text{ particles of gas A}}{\text{mass of } n \text{ particles of gas B}} = \frac{\text{mass of 1 particles of gas A}}{\text{mass of 1 particles of gas B}} = \frac{\text{mass of 1 mole of gas A}}{\text{mass of 1 mole of gas B}}$$

Weighing a gas is more complicated than weighing a small solid object for many reasons. First, small volumes of gas have small masses. Therefore, even tiny errors impact the result. Also, objects float if the air that is displaced by the object weighs more than the object displacing it. This phenomenon, buoyancy, is easily seen in a helium-filled balloon. The balloon floats in air, and if a helium balloon was placed on a balance, it would appear to weigh nothing! Objects that do not float are still affected by an upward force equal to the mass of air they displace and proportional to the volume of air that is displaced. For example, although your mass, the amount of matter you contain, is the same in air or in a vacuum, you would appear to weigh about 100 grams (about 1/4 lb) less in air then you would weigh in a vacuum, depending on your volume. To get an accurate weight of an object, the mass of the air displaced by the object must be added to the weight of the object in air. For solid objects, the correction is so small that it can be ignored in most cases. However, when weighing gases, this buoyancy effect is significant.

In this investigation, you will weigh equal volumes of different gases. You will then correct the data for buoyancy to determine the mass of each gas sample. Finally, you will use the relationships between mass and volume of gases to relate the corrected masses to a standard reference substance (oxygen, O_2). You will then be able to determine values for the molar masses of the gases that you have investigated.

Making Predictions

- Rank the following gases from lowest to highest molar mass: oxygen, methane, carbon dioxide, argon, and nitrogen. Explain why you put them in this order.

- After reading *Gathering Evidence*, prepare a data sheet that starts with the predicted ranking and includes space for observations and measurements (including temperature and pressure in the laboratory). Also, make a table for collection of data for each of the listed gases and an unknown. Some possible data for the table might include weight of gas assembly, measured mass of gas, mass of air displaced, corrected mass of gas, calculated (or investigation) molar mass, and accepted molar mass.

- Using values for atomic weights from the periodic table, calculate the molar mass of each gas you will measure in the investigation to the nearest whole number. Enter these values into the table on your data sheet as accepted molar masses for the gases.

Gathering Evidence

Overview of the Investigation

1. Prepare a plastic bag to use as a gas container.
2. Fill the container with several different gas samples, including one unknown gas.
3. Weigh the container filled with each of the gases.
4. Determine the volume of the container.
5. Determine the mass of 1 liter of air at room temperature and pressure.
6. Calculate the measured and corrected mass of each gas.
7. Calculate the molecular mass and volume of each gas.
8. From the data, propose a possible formula for the unknown gas.

Part I. Weighing the Gases

Notes about weighing: The most critical measurements in this investigation are masses of the gas container plus gases that are obtained by means of a laboratory balance. The use of a balance is explained in detail in the *Laboratory Methods* section, and your instructor will explain how to use the particular balances available in your lab.

Remember, the balance must always read exactly zero when nothing is on the balance pan. It is a good idea to check it and tare it before each measurement. Also, the readings should be recorded to at least the nearest 0.01 g. For this measurement, you will record the digit two places to the right of the decimal point, even if this digit is zero. If your balance shows a third decimal place, record it.

CAUTION! A single drop of water in the bag will add enough mass to impact the results. Make sure the bag is dry at the start and weigh all of the gases before you add water to determine the volume of the bag.

1. Gather materials for the weighing device, including a plastic bag, a one-hole rubber stopper with a medicine dropper stuck through the hole, a cork ring with a ring that fits the rubber stopper, and a dropper cap (Figure 10.1).

2. Push the open end of the plastic bag through the center of the cork ring. Open the bag, and push the rubber stopper firmly into the open bag so that the bag is tightly caught between the stopper and the ring. Make sure there are no leaks in the system. To do this, inflate the bag with one of the gases, put the rubber cap over the hole in the medicine dropper, and gently squeeze the bag. No air should escape.

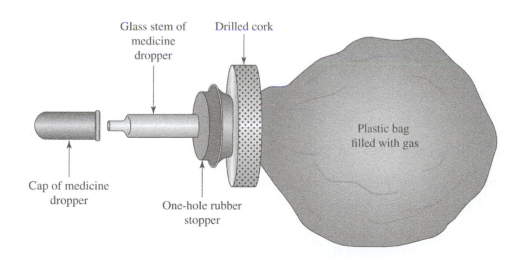

Figure 10.1. Diagram of the assembled system.

3. Take the rubber cap off of the medicine dropper and press all air out of the bag. Be sure to completely flatten the bag to get all the air out so that you are weighing a completely empty bag. Once the air is out of the bag, replace the cap on the medicine dropper.

4. Tare the balance so that it reads 0.00 g. Weigh the bag assembly and record the mass to the nearest 0.01 gram.

5. Remove the cap and connect the bag to a source of gas using the medicine dropper. Hold the bag by the stopper and completely fill the bag with the first gas you are investigating.

7. Do not squeeze the bag, but allow any excess gas to escape so the gas in the bag will be at the same pressure as the room. Replace the rubber cap on the dropper.

8. Weigh the bag assembly, which contains the gas (at room temperature and atmospheric pressure), and record your measurement. Make sure that the bag assembly is completely on the balance pan and does not touch any part of the balance frame or housing.

9. Remove the cap from the dropper and press all of the sample gas out of the bag. Repeat the procedure (Steps 5–8) for the other gases available in the lab, including the unknown gas assigned to your team.

Part II. Finding the Volume of the Bag

1. After you have weighed all the available gases, fill the bag with air at room pressure and attach a rubber hose to the medicine dropper (Figure 10.2).

2. Fill a large pan or the sink with enough water to cover the mouth of a large inverted bottle of water.

3. Completely fill a large bottle with water. Do not leave any air space in the bottle.

4. Cap the bottle and invert it. Place the inverted bottle in the sink or pan of water so that the mouth is below the surface of the water. Make sure there are no air bubbles in the bottle.

5. While the bottle opening is underwater, remove the cap from the bottle.

6. With the help of your lab partner, insert the free end of the rubber hose into the mouth of the bottle, as shown in the diagram.

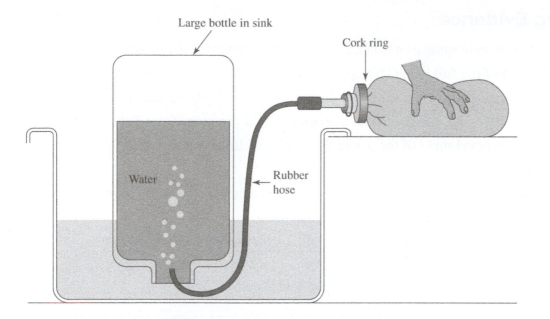

Figure 10.2. Diagram of the setup used to determine the volume of the plastic bag.

7. Gently squeeze the air out of the plastic bag. As you do so, bubbles of air will enter the bottle and force some of the water out of the bottle.
8. When the bag is empty, keep the bottle inverted with the mouth under water while you replace and tighten the screw cap.
9. Remove the bottle from the sink or pan, place it in an upright position, and remove the cap.
10. Fill a 1000-mL (1-L) graduated cylinder with water so the meniscus is exactly at the 1000-mL line. Slowly pour water from the cylinder into the open bottle until the bottle is completely filled. Record the volume of water left in the graduated cylinder. Subtract the volume of water remaining in the graduated cylinder from 1000 mL and record this number on your data sheet. This value is the volume of water that refills the bottle. Since this amount of water was displaced by the air in the bag, the value also is equal to the volume of air that was originally in the bag.
11. Record the room temperature and air pressure. Your instructor may provide these values for the entire class, or you may be asked to measure them with a thermometer and a barometer.
12. Use Table 10.1 to determine the mass of 1 L of air at the temperature and air pressure of the room. Record this value on your data sheet. *Note:* In the chart, air pressure is given in mmHg, the height of a column of liquid mercury in a mercury barometer. For comparison, standard atmospheric pressure at sea level is 760 mmHg.

Analyzing Evidence

1. Calculate the apparent mass of gas in the bag by subtracting the weight of the empty bag from the weight of the full bag. (This apparent mass may be less than zero. How can that be?)

2. Calculate the mass of air displaced by the bag of gas by multiplying the volume of the bag (in liters) by the mass of 1 L of air at room temperature and pressure (from Table 10.1).

3. Calculate the corrected mass of the gas in the bag by adding the mass of the displaced air to the apparent mass of the gas.

4. Use the defined molar mass of oxygen (32 g/mol) to calculate the investigation molar masses of the other gases, by means of the following equation.

$$\frac{\text{molar mass of gas A}}{\text{molar mass of oxygen}} = \frac{x \text{ g/mol}}{32 \text{ g/mol}} = \frac{\text{measured mass of gas A}}{\text{measured mass of oxygen}}$$

5. Record all of these results in the table on your data sheet.

Table 10.1. The mass (in grams) of 1 liter of air at different temperatures and pressures.

Pressure	Temperature			
(mm Hg)	15 °C	20 °C	25 °C	30 °C
600	0.97	0.95	0.94	0.92
610	0.99	0.97	0.96	0.94
620	1.00	0.98	0.97	0.95
630	1.02	1.00	0.99	0.97
640	1.03	1.01	1.00	0.98
650	1.05	1.03	1.02	1.00
660	1.07	1.05	1.03	1.01
670	1.08	1.06	1.05	1.03
680	1.10	1.08	1.06	1.04
690	1.11	1.09	1.08	1.06
700	1.13	1.11	1.09	1.07
710	1.14	1.12	1.11	1.09
720	1.16	1.14	1.12	1.10
730	1.18	1.16	1.14	1.12
740	1.19	1.17	1.15	1.13
750	1.21	1.19	1.17	1.15
760	1.22	1.20	1.18	1.16

Interpreting Evidence

1. Using the molar masses you calculated from your investigation results, rank the following gases from lowest to highest molar mass: oxygen, methane, carbon dioxide, argon, and nitrogen. Compare this ranking to your predicted ranking. Explain any differences and use evidence to support the ranking you think is correct.

2. Weighing a gas is more complicated than weighing a small solid object. Describe how the following errors in the measurement of mass would change the data. Would your mass be too low or too high? Why?

   a. Not emptying the bag completely before weighing the assembled system.

   b. Having the bag resting on part of the balance while measuring the weight of a gas.

   c. Not correcting for buoyancy.

3. Compare your values for the molar mass of each gas with the accepted value. Do your results support the hypothesis that equal volumes of gases have equal numbers of particles? Briefly explain the basis for your answer.

4. What are some likely sources of error that could account for the difference between measured and accepted values for molar mass? Indicate whether each one would make your answer too low or too high. (*e.g.*, How would the results of your investigation change if the temperature of one of the gas samples changed before it could be weighed?)

Making Claims

What can you claim about the molar masses of different gases? Is this a good method for measuring molar mass?

Reflecting on the Investigation

1. One of the first ways that was used to determine the molar mass of a liquid compound utilized the following procedure: The liquid was placed in a glass container, and the container was heated until all of the liquid had boiled away. As a consequence, the air originally in the container was totally displaced by gaseous molecules of the unknown compound. The container was sealed (while still hot), then cooled and weighed. What information do you think was needed to complete the determination of the molar mass of the liquid?

2. Why can't we weigh a single atom or molecule? Explain why finding the mass of a mole of particles is a good substitute.

What is a mole?

The concept of the mole is one that students often struggle with. Despite an instructor's attempts to relate it to a dozen eggs or a year's worth of days, students are often left without a clear understanding of what it means. These simple and fun activities may help clarify what a mole is and how it is used by chemists in calculations.

Activity 1 – Brick chemistry

For this activity, you will need an analytical balance and a collection of interlocking building bricks, such as Lego™.

1. Take a structure built completely out of one size of interlocking building bricks, such as 2×4 or 1×2 bricks. How can you figure out the mass of one brick without taking the structure apart?

2. Once you know the mass of one brick of a certain size, take a bag containing many bricks of that size. How can you measure how many bricks are in the bag without removing them from the bag and counting them? Can you calculate how many tens of bricks you have? How about how many dozens of bricks?

Activity 2 – How many atoms?

For this activity you will need an analytical balance and some items made of a single element, such as an aluminum pan, fishing weights made of tungsten, gold jewelry, or an artist's stick of graphite.

1. Find and record the mass of each object.

2. Use the periodic table to find the atomic mass of the element found in each object. Use the atomic mass to calculate the moles of atoms within each object.

3. Use Avogadro's number to calculate the number of atoms within each object. Why do you think chemists often prefer to use moles rather than number of atoms in their calculations?

4. For a more challenging activity, consider the purity of the atoms within each object. For instance, 18-carat gold is 75% gold and 25% other metals, usually copper or silver. How would you find the number of moles of gold in 18-carat gold?

Verifying Molar Ratios in Chemical Reactions

Asking Questions

- What is the relationship between amounts of reactants and amounts of products?
- How do the concepts of moles and molar ratios help define the changes in chemical reactions?
- How can we use the masses of reactants and products to verify the molar ratios in a balanced equation?

Preparing to Investigate

In this investigation, you will collect quantitative measurements for a reaction between sodium bicarbonate, $NaHCO_3$, and hydrochloric acid, HCl. Sodium bicarbonate is also known as sodium hydrogen carbonate, but you may be familiar with it as baking soda. The reaction of sodium bicarbonate with hydrochloric acid produces table salt (sodium chloride, NaCl), water, and carbon dioxide. The balanced chemical equation is shown here.

$$NaHCO_3\ (s) + HCl\ (aq) \rightarrow NaCl\ (aq) + H_2O\ (l) + CO_2\ (g)$$

You will determine by investigation how many moles of salt (NaCl) are formed from a known number of moles of $NaHCO_3$. During chemical reactions, substances combine with each other in a definite proportion by mass, meaning that only a certain amount of one reactant will react with a given amount of another reactant. The amounts of reactant species can be expressed in a variety of ways: grams, pounds, tons, or liters, for example. However, no matter what units are used, they all relate to the ratio of *moles* of one species that react with a certain number of *moles* of another species. If you are unclear about the concept and definition of a **chemical mole**, you should review the discussion of moles in Chapter 4 of *Chemistry in Context*.

On the basis of the chemical equation shown above, the molar ratio of reactant to product is 1:1. This means that for every mole of sodium bicarbonate that reacts with hydrochloric acid, a mole of sodium chloride will be formed. In this investigation, you will verify this ratio by determining the mass of $NaHCO_3$ used and the mass of NaCl formed by weighing samples on a balance. The masses can be converted to moles, and the molar ratio can be verified.

You can also repeat the process with a different reagent to verify the molar ratio for the production of sodium chloride from sodium carbonate (Na_2CO_3). To further improve the accuracy of the results, each group will collect multiple data points for each compound, and the class data will be combined and analyzed.

Making Predictions

- Write out the balanced equations for the chemical reactions between sodium bicarbonate and HCl and sodium carbonate and HCl. Without doing any calculations, predict the molar ratios of reactants and products with each reaction and explain why you think these ratios reflect the chemical reactions.

- After reading *Gathering Evidence*, prepare a data sheet that starts with the equations and predicted ratios. Make a table for collection of data that includes a column for each of the three trials and rows for the mass of the tube, mass of the tube with compound, mass of the compound, molar mass of the compound, moles of the compound, mass of the tube with product, mass of the product, moles of the product, and molar ratio. If you are investigating both sodium bicarbonate and sodium carbonate, make sure you have enough entries in your data table to record data for both reactions.

- Using values for atomic weights from the periodic table, calculate the molar mass of each compound (reactants and product) you will measure in the investigation to the nearest whole number. Enter these values into the table on your data sheet as accepted molar masses for the compounds.

Gathering Evidence

Overview of the Investigation

1. Determine the mass of sodium bicarbonate you will use.
2. React the sodium bicarbonate with 10% hydrochloric acid.
3. Determine the weight of sodium chloride produced.
4. Calculate the ratio of moles of NaCl formed to moles of $NaHCO_3$ used.
5. Repeat process using sodium carbonate instead of sodium bicarbonate.

Conducting the Reaction and Measuring Reactant and Product Masses

Notes about weighing: This investigation requires careful weighing of test tubes on a laboratory balance. The technique of measuring mass is described in detail in the *Laboratory Methods* section of this book, and your instructor will explain how to use the balances available in your lab. Remember, the balance must always read exactly zero (0.00 g) when nothing is on the balance pan. It is a good idea to check it and tare it before each measurement.

1. Obtain three test tubes that are clean and completely dry. Label them at the top of the tubes with A, B, and C. Add a small boiling chip to each test tube.

2. Take the test tubes and your data sheet to one of the laboratory balances. Weigh each of the test tubes (including the boiling chip) to the nearest 0.01 g and record the masses in your data table. For accuracy, be sure to record the values to the second decimal place, even if the second decimal place is a zero (*e.g.*, 18.10 g).

3. To each test tube, add just enough $NaHCO_3$ to fill the curved bottom of the tube.

4. Weigh each test tube again with its contents to the nearest 0.01 g and record the mass in your data table. *Note:* The masses of the three solid samples do not need to be the same. Typical masses of added $NaHCO_3$ should be between 0.30 to 0.70 g.

5. At your lab bench, fill a plastic pipet with 10% hydrochloric acid solution. Add the acid dropwise to tube A. Let each drop of the liquid run down the wall of the test tube and, after each drop reaches the bottom, gently tap the test tube. Continue to add acid slowly until all of the solid has reacted. Keep in mind that it is important to add only the *minimum* amount of acid needed to react with the solid. Put the tube aside for Step 7.

 CAUTION! 10% HCl is corrosive to the skin and other materials. Avoid spilling it on yourself, your partner, or your work space. If any comes into contact with your skin, immediately rinse with water and inform your instructor.

6. Repeat Step 5 with each of the remaining test tubes (B and C).

7. To isolate the NaCl, you must allow the water to evaporate. You will need to heat the tubes to dryness over a Bunsen burner.

 STOP! No flammable chemicals should be in the vicinity of the Bunsen burner. Long hair should be tied back and loose sleeves on clothing should be secured.

Note: Too rapid heating of the tube, especially if it is held in an upright position, will cause the hot contents to splash out of the tube and will necessitate starting over with a fresh sample. Boiling chips should help to produce smooth boiling. Your instructor may provide additional advice on how to minimize the problem.

8. As in Figure 11.1, hold the test tube at an angle, and point it away from you and anyone else in the immediate vicinity. Gently heat the tube and its contents over the flame. You want to evaporate the water in the tube without its contents boiling over or splattering.

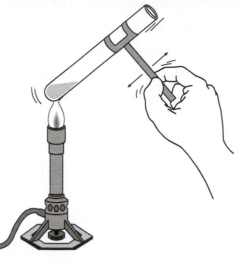

Figure 11.1. The correct way to heat a test tube over a Bunsen burner.

9. Continue heating until *all* of the liquid has vaporized and solid only NaCl remains. (It is crucial to the success of this investigation to be sure that no liquid water has condensed on the *upper* part of the tube.)

10. Remove the tube from the flame and test for the evolution of water vapor from the test tube by inverting a clean, dry test tube over the upright mouth of the test tube. If condensation occurs in the cold test tube, continue the heating and testing process until no condensation occurs. Then set the test tube with its dried contents aside to cool.

11. Repeat Steps 7 through 10 with test tubes B and C.

12. Allow the three test tubes to cool (at least 5 minutes), *check to be sure there are no water droplets left*, and then weigh each with its contents. Record the masses in your data table.

13. If time permits, confirm that the tubes were fully dried by reheating them for 1–2 minutes, cooling for 5 minutes, and reweighing. Record the second mass for each. If they were dry the first time, there should be a negligible change in mass.

14. After completing the study with sodium bicarbonate, you may try a similar study with sodium carbonate (Na_2CO_3) to see how the results differ. If so, repeat Steps 1 through 13 with clean, dry test tubes. If you do these trials, you will need to have additional space in your data table for the masses and calculations.

Clean up

Dispose of the test tube contents as your instructor directs you. Do not put anything down the sink drain without being told that it is permitted. Thoroughly wash the test tubes, rinse with deionized water, and leave them upside down to drain.

Analyzing Evidence

1. For each test tube, subtract the mass of the test tube from the test tube with added $NaHCO_3$ to determine the mass of $NaHCO_3$ used.
2. Use the molar mass of $NaHCO_3$ that you previously calculated and the masses from Step 1 to calculate the number of moles of $NaHCO_3$ you used in each trial.
3. For each test tube, subtract the mass of the tube from the final mass of the test tube with product to obtain the mass of $NaCl$ formed by the reaction.
4. Calculate the moles of $NaCl$ produced.
5. Calculate the ratio of moles $NaCl$ to moles $NaHCO_3$. This ratio should be recorded to two decimal places (*i.e.*, two digits after the decimal point).
6. Calculate the average of your three investigation results for this ratio and record it to two decimal places.
7. Repeat calculations for Na_2CO_3, if applicable.

Interpreting Evidence

1. Combine your data with those of the rest of the class. Do your investigation results, or those of the class, agree with the balanced equation? Discuss your answer. If you think any results should be excluded from calculating the average class ratio, explain why and calculate a revised average ratio.

2. Suggest two possible sources of error in this investigation (do not include weighing errors). Indicate whether each error would increase or decrease the investigation value for the mole ratio.

Making Claims

What can you claim about chemical equations and molar ratios of reactants and products? What evidence from this investigation can you use to defend your claim?

Reflecting on the Investigation

Balance each equation below. It is a mechanism that shows how ozone (O_3) forms in the upper atmosphere. Use this process to determine how many moles of O_2 are required to form one mole of O_3. Defend your answer using evidence from the investigation.

a. First, nitrogen and oxygen gas react to form nitrogen monoxide.

$$N_2 + O_2 \rightarrow NO$$

b. Then, nitrogen monoxide reacts with more oxygen to form nitrogen dioxide.

$$NO + O_2 \rightarrow NO_2$$

c. When nitrogen dioxide is struck by UV light, it decomposes to form nitrogen monoxide and a highly reactive oxygen atom.

$$NO_2 + UV \text{ light} \rightarrow NO + O$$

d. Finally, the oxygen atom reacts with diatomic oxygen to form ozone.

$$O + O_2 \rightarrow O_3$$

Hot Stuff: An Energy Conservation Problem

Asking Questions

- What are some ways to measure properties that you can't see?

- How do scientists approach problems that don't have obvious solutions?

- What data is important when defending a method for solving a problem?

Preparing to Investigate

This laboratory investigation is a departure from the usual investigation. It simulates the kind of problem solving that takes place in a scientific laboratory. There are no instructions, procedures, or data sheets. There is simply a problem to solve that requires reasoning skills and the application of previous experience or knowledge.

A variety of materials and equipment will be available that you can use to solve this problem. At a minimum you should have access to graduated cylinders, burets, beakers, flasks, plasticware, disposable cups, stirrers, test tubes, plastic pipets, a balance, and a thermometer that only reads to 40 °C.

Gathering Evidence

Organizing the Team

Your team must devise a method for solving the problem and obtain a numerical answer. A team will typically consist of three or four members. Although all members of the team should be involved in the entire process, your investigation will be more efficient if you take a few minutes at the beginning to be sure each person has a clear role in the investigation. Possible positions you might want to assign to members of the group include a team leader, a "go-fer," an experimentalist, a report writer, and a recorder. It especially will be important to keep track of ideas, data, and investigation procedures during the investigation. When finished, your instructor will lead a class discussion that will allow teams to compare methods and answers.

Measuring Water Temperature

As an energy conservation method, you decide to turn down your home water heater so that it only heats water to 55 °C. (It previously heated the water to between 60 °C and 70 °C.)

Unfortunately, when you decide to measure the temperature of the water in your water heater, the only thermometer available has a maximum temperature of 40 °C.

In the laboratory, there is a simulated water heater, which consists of a large coffee pot filled with hot water. Each team will have available a thermometer that has the graduations above 40 °C covered so that you cannot see them. Using your 40 °C thermometer and materials available to you in the lab, devise a way to measure the temperature of the water in the coffee pot.

Optional extension: Your instructor may ask you to propose two or three *different* methods for solving this problem, then test each of them, decide which gives the most reliable answer, and explain why.

Analyzing and Interpreting Evidence

Your team should be prepared to defend your answer and the method you used to obtain it. Therefore, it is important for your team to keep a complete record of everything you do and the numerical data you obtain. It also is important that you present the data to provide evidence that your method solves the presented problem. In addition to an oral report to the class, your group should prepare a one-page written report of your investigation. Your instructor will specify what should be included in the report

Making Claims

What can you claim about your method for measuring the temperature of the hot water? Include these claims in your oral and written reports.

Reflecting on the Investigation

When everyone is finished, your class will assemble to hear a report from each team about the method they used and their results. Take notes and identify the strengths and weaknesses of each team's approach.

Comparing the Energy Content of Fuels

Asking Questions

- How can the energy content of fuels be measured?

- Are some fuels more efficient than others?

- Which fuel do you think would release the most energy upon burning, ethane (C_2H_6) or ethanol (C_2H_6O)? Why?

- What is the relationship between the presence of oxygen in a fuel and its energy content?

- Why can't we combust carbon dioxide or water ? What prevents argon from combusting?

Preparing to Investigate

In this investigation, you will investigate the energy content of several fuels by using them to heat water. The data that you and your classmates obtain will enable you to compare some fuels to see which ones provide more energy for a given mass of fuel burned. In particular, you will be able to determine how the energy released by burning **hydrocarbons**, compounds that contain only hydrogen and carbon, compares to the amount of energy released by burning oxygenated fuels such as ethanol, a renewable **biofuel** derived from biological rather than geological sources.

Fuels burn and give off energy. They do so by combining with oxygen to form compounds of lower energy, typically carbon dioxide and water. For example, here is a chemical equation that represents the **combustion** of methane, CH_4, to produce CO_2 and H_2O.

$$CH_4 + 2\,O_2 \longrightarrow CO_2 + 2\,H_2O$$

Methane burns cleanly. We call this *complete combustion* because the two products, CO_2 and H_2O, are not flammable; that is, they cannot burn any further. Other hydrocarbons, such as propane (C_3H_8), which is used in barbecue grills, and butane (C_4H_{10}), which is used in camping stoves, will burn to produce the same products: CO_2 and H_2O.

Other fuels, however, may burn incompletely unless supplied with plenty of oxygen. In addition to carbon dioxide, the combustion reaction also produces carbon monoxide and/or soot. For example, some hydrocarbons, including candles, burn with a sooty flame. When either soot or carbon monoxide or both are formed as products, this is called *incomplete combustion*. Here is the chemical reaction that shows how hexane (a hydrocarbon, C_6H_{14}) burns to produce carbon monoxide.

$$C_6H_{14} + 2\,O_2 \longrightarrow 6\,CO + 7\,H_2O$$

Some fuels such as ethanol (C_2H_5OH) contain oxygen in addition to carbon and hydrogen. In essence, they are hydrocarbons that have already partially reacted with oxygen:

$$2\ C_2H_6 + O_2 \longrightarrow 2\ C_2H_5OH$$

The energy when fuels are burned is given off primarily in the form of heat, with some light as well. Although sometimes quite bright, the light emitted by a flame or fire is hard to quantify. In contrast, it is straightforward to measure the heat released when a fuel burns. For these two reasons, you will evaluate the energy content of fuels in this investigation by the heat they give off rather than by the light.

But how do you measure the heat? Rather than measuring it directly, you will measure the heat released to increase the temperature of a known quantity of water. From this, you can calculate the heat released by burning a particular fuel.

Some of the fuels you will test include alcohols such as methanol (CH_3OH), ethanol (C_2H_5OH), iso-propanol (C_3H_7OH), or butanol (C_4H_9OH). Others will be hydrocarbons, such as lamp oil or candle wax. The latter are actually mixtures of hydrocarbons, but we will approximate their compositions as $C_{12}H_{26}$ (lamp oil) and $C_{40}H_{82}$ (candle wax).

You will heat water by burning a known amount of a fuel. It takes exactly 1 calorie (cal) of heat to raise the temperature of 1 gram of water by 1°C. Therefore, if you know the mass of water and how many degrees the temperature increases, then the total amount of heat absorbed by the water can be calculated:

$$\text{heat absorbed (calories)} = \text{mass of water (grams)} \times \text{temp. change (°C)} \times 1.00\ \text{cal/g•°C}$$
$$= m \times \Delta T \times 1.00\ \text{cal/g•°C}$$

In this equation, ΔT represents the change of temperature, and the last term, 1.00 cal/g•°C, is the **specific heat** of water. Confirm that combining and canceling the units on the right side of the equation will leave only calories. Theoretically, the amount of heat liberated by the burning fuel should equal the heat absorbed by the water, but in practice, some of the heat will be lost to the surroundings.

Making Predictions

- List the fuels you will study in this investigation and predict their relative energy outputs.
- After reading *Gathering Evidence* and the appropriate parts of *Laboratory Methods*, prepare a data table to record your masses, temperatures, and heat absorption results.

Gathering Evidence

Overview of the Investigation

1. Assemble the apparatus and obtain a burner containing a known fuel.

2. Add a measured volume of water to the can and determine the mass of the water.

3. Weigh the burner with the fuel.

4. Record the initial temperature of the water.

5. Light the burner and heat the water until the temperature increases about 20 ° C.

6. Extinguish the burner and record the highest temperature of the water.

7. Weigh the burner again to find the mass of fuel used.

8. Repeat with two or more additional trials.

9. For each trial, calculate the amount of heat released per gram of fuel burned.

Part I. Measuring the Energy Content of Different Fuels

The detailed procedure for **calorimetry** is described in the *Laboratory Methods* section of this lab manual, and you should read that section before proceeding.

Note: The success of this investigation depends heavily on the accuracy of your measurements. Perform all weighing operations on a laboratory balance. You should read the section on measuring mass in *Laboratory Methods,* and your instructor will explain the use of the particular balances in your lab. Before each measurement, it is important to check that the balance reads *zero* with nothing on the balance pan. Record your measurements to at least 0.01 g.

Your instructor will tell you which fuel or fuels you are to investigate. You may be asked to do one trial with each of three different fuels. Alternatively, you may be asked to do several trials with a single fuel.

 CAUTION! You need to be constantly aware when you and other students are working with flammable solvents and open flames. Handle the fuel burners very carefully. Before starting the investigation, be sure you know where a fire extinguisher is located in your laboratory. As with any investigation involving open flames, long hair must be tied back and loose sleeves should be secured. Wear eye protection at all times.

Part II. Optional Extensions

1. **Investigate the effects of changes in procedure**. These might include adding some nonflammable insulation around the can, adding a cylindrical shield of aluminum foil

Clean up

Discard or recycle the contents of your test tubes as directed by your instructor. Clean your thermometer with soap and water to remove all traces of the oil.

Part IV. Measuring the Heat Content of Biodiesel

STOP! Handle the fuel burners very carefully. Before starting the investigation, be sure you know where a fire extinguisher is located in your laboratory. As with any investigation involving open flames, long hair must be tied back and loose sleeves should be secured. Wear eye protection at all times.

1. Prepare a burner for your biodiesel. Obtain a 20-mL beaker, a candle wick, and a paper clip. The end of the candle wick should be wrapped around one wire of the paper clip. Spread the paper clip out slightly so that it makes a stable bottom support for the wick. The wire inside the wick will allow it to stand vertically. Put the paper clip and wick in your beaker. The wick should extend to the top of the beaker. Fill the beaker with at least 15 mL of your biodiesel fuel.

2. The wick should stand freely in the middle of the biodiesel. Use matches to light the wick and ensure that your burner is working well. Once you know that your burner is working, extinguish its flame.

3. Measure the energy content of your biodiesel by following the calorimetry procedure given in the *Laboratory Methods* section. Your apparatus will look similar to that in Figure 0.11 of that section, except that the burner in this case is your beaker filled with biodiesel. Repeat the measurement twice.

Analyzing Evidence

1. Look carefully at the data from each of the sets of viscosity estimates to see whether the results show consistency or whether any one result in a given set should be eliminated because it appears to be an outlier. If so, draw a single line through the result in your data table and make a note that you have not included it in your analysis.

2. Calculate the average amount of time that it took for each liquid to drain.

3. Calculate the heat released by your biodiesel for each trial and find the average of your trials.

Interpreting Evidence

Although it was only possible for you to do a few trials, it is desirable to assemble more data so that better comparisons of biodiesel fuels can be made. Your instructor will tell you how to report your results to the rest of the class, so that you can compute averages and make comparisons. The following questions should help to focus your interpretation of the results.

1. You made a prediction as to which layer in your centrifuge tube would be on top. Which layer actually was on top? Was your prediction correct? If yes, what basis did you use for the prediction? If no, what error did you make?

2. To get an accurate measurement of the biodiesel fuel energy content, all the heat released when it burns must be transferred to the water in the can. We never get it all. Where else could it go? As a result of these losses, do you expect your calculated heat content to be a little too low or a little too high when compared to the actual heat content of the biodiesel? Explain.

3. If your class prepared biodiesel from several different oils, prepare a small table listing the drip time and the heat content for each type of biodiesel. Now rank the fuels in order from lowest to highest energy content. Compare your ordered list to Figure 5.6 in the *Chemistry in Context* textbook. What do you notice?

Making Claims

What can you claim about biodiesel as an energy source?

Reflecting on the Investigation

1. You observed in this investigation that biodiesel and water do not mix. Give two other examples of some type of oil not mixing with water.

2. Predict the density of gasoline, given what you know about the density of water and given what you observed in this investigation. Then look up the value to see how close you came.

NOTES

Measuring Radon in Air

Asking Questions

- What is the relationship between alpha particles and helium nuclei?
- What property of radon gas can be used to detect it?
- What factors impact radon levels in buildings?
- Why is it important to measure radon concentrations in our living environment?
- How can high levels of radon be mitigated?

Preparing to Investigate

Radon is a radioactive gas that is a potentially serious indoor health hazard in some geographic areas (see Chapter 6 of *Chemistry in Context*). Radon-222, the longest-lived isotope of radon, is an alpha-emitter with a half-life of 3.8 days. The atomic number of radon is 86, which places it in Group 18 at the far right-hand side of the periodic table. The elements in Group 18 are chemically inert gases that generally do not form molecules.

When radon undergoes alpha decay, its radiation is measured in picocuries (pCi). Usually, it's monitored in a specific volume of air, so the final unit is pCi/L.

Radon is formed as an intermediate product in the radioactive decay of uranium-238. Uranium is widely distributed on the surface of our planet. For example, granite rocks contain about 4 ppm of uranium. When radon is produced from uranium, it passes through fissures in rocks and soils without reacting and enters homes through cracks or other holes in the foundations. The amount of radon found in homes varies from one part of the country to another, and is typically a more serious concern in houses with basements. National and state maps of predicted radon concentrations in homes can be found on the U.S. Environmental Protection Agency (EPA) website.

In this investigation, you will use a simple radon detector to measure the radon concentration in a location of your choosing. A sampling period of at least 4 weeks is necessary to obtain reliable data. The disk in the detector consists of a high-clarity polymer often used in eyeglasses that is known as CR-39. The full chemical name for CR-39 is poly[ethylene glycol bis(allyl carbonate)]. When alpha particles emitted from radon penetrate CR-39, they cause damage in the plastic, probably due to disruption of the polymer chain along the path of penetration. Although the damage is not visible to the eye, treating the disk with sodium hydroxide, NaOH, reveals it. Sodium hydroxide etches the sample in the damaged regions. The etched regions show up as "tracks" when viewed under a microscope and the number of tracks in a given area of the disk can be used to estimate the radon level at the sampled location. Although beta and gamma radiation also penetrate the plastic, these rays mostly pass through without causing damage.

Making Predictions

Look at the maps on the EPA website. Do you predict that radon will be found in the area where you placed your sampling disk? Create a data sheet that will hold your data and observations for this investigation.

Gathering Evidence

Overview of the Investigation
1. Place a detector disk in an undisturbed location for at least 4 weeks.
2. Etch the disk to produce visible alpha tracks.
3. Count the tracks under a microscope.
4. Calculate the concentration of radon in air by comparing it to a control disk.

Part I. Sampling the Air

Start the air sampling early in the semester, as directed by your instructor. A long sampling period, preferably for *at least* 4 weeks, is necessary. Select a location where the detector can remain undisturbed and where you can retrieve the detector prior to the scheduled lab time for this study. *Note:* Plan this carefully, perhaps adding a label to explain what the item is. Custodians or other persons may unknowingly remove the disk when cleaning up an area.

The class results will be more interesting if students select many locations, including on- and off-campus, outdoors, and inside buildings at various floor levels and with different ventilation or air circulation systems. One possibility is to use your own or someone else's home, either taking the detector device there yourself or mailing it to a friend or relative with instructions. In some parts of the country, basements of houses are particularly prone to high radon levels.

A. Materials
1. Piece of CR-39.
2. Small plastic cup, preferably with lid.
3. Rectangle of cardstock (size depends on the cup being used).
4. Piece of transparent adhesive tape.
5. Piece of thin, single-layer toilet or tissue paper (to protect the CR-39 from dust).

B. Procedure for Use at Sampling Location

1. If the cup has a lid, cut a large circular hole in the lid.
2. Remove the plastic film from the marked side of the CR-39.
3. Make a loop of tape with the sticky side out. Stick the tape to the side of the CR-39 that still has the plastic film coating in place. Stick the CR-39 to the middle of the piece of cardstock.
4. Bend the cardstock into an inverted U that will support the CR-39 near the top of the cup. Put the cardstock with the CR-39 into the plastic cup. Cover the opening of the plastic cup with the toilet or tissue paper, and then snap the lid onto the cup. If the cup does not have a snap-on lid, use a rubber band to hold the tissue paper in place.
5. Place the detector in the chosen location, and record the location and start date and time.

    Note: It is important that you accurately record the location for the detector disk (describing it in detail) and the start and stop times (dates and time of day).
6. At the end of the air-sampling period, record the ending date and time. Store the card with the attached disk in a labeled envelope until time for the laboratory work.

Part II. Etching the Disk

Remember to bring your exposed disk to your laboratory class!

1. Remove the disk from the cardstock. You already peeled off one of the polyethylene films from the disk. Now peel off the other (from the back of the disk). Gently slip the ring of an "etch clamp" over the top of the disk. *CAUTION:* The disk is fragile and can break easily. Fasten the ring onto a paper clip that has been bent so there is a hook at each end (Figure 15.1).

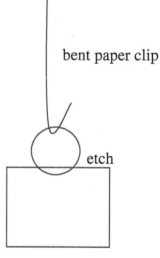

bent paper clip

etch

CR-39

Figure 15.1.

Investigation setup for etching.

2. Hook the disk inside a test tube and use a pipet to add enough 6-M sodium hydroxide (NaOH) to cover the disk in the tube. Do not submerge any more of the paper clip than is necessary because the coating on the paper clip will react with NaOH.

    CAUTION! Hot 6-M NaOH is extremely corrosive to skin and to clothing. Wear goggles at all times. Rubber gloves are strongly recommended.

3. Heat a beaker of water on a hot plate to a boil. Carefully place the test tube containing the disk and NaOH solution in the boiling water bath and heat it for 40 minutes. At the end of 40 minutes, remove the disk and rinse thoroughly with lots of tap water.

Clean up

When finished, do NOT pour the NaOH solution down the drain. Pour it into an approved waste container provided by the instructor.

Part III. Counting Alpha Tracks

The measuring process will be completed using a microscope. Your instructor will show you how to operate the microscope that you will be using.

A. Calibrating and focusing the microscope

1. Determine the area (in square centimeters) of the field of view of your microscope at low power (10x) by looking through the microscope at a ruler. Count the number of millimeter divisions on the ruler that you can see across the widest portion of the field of view and record the diameter in millimeters. Divide by 10 to find the diameter in centimeters. The radius, r, is half of the diameter, and the area, A, can be calculated by using the formula $A = \pi r^2$.

2. Place an etched disk on a microscope slide under the microscope. Adjust the focus until you can clearly see the etched disk. Sometimes it is easiest to focus on the edge of the disk and then move it to the part of the disk you wish to observe.

3. Practice looking at alpha tracks on a previously exposed and etched disk. This may be a disk donated by a student in a previous class, or it may be one prepared by your instructor using a radioactive source (in which case, your own disk likely will have fewer tracks).

B. Counting alpha tracks

1. Look at the two pictures of alpha tracks at different magnifications (Figure 15.2). The left photo (Figure 15.2 a) looks more like what you will see. At higher magnification, you can see tracks of different shapes (Figure 15.2 b). The various shapes are the result of how the alpha particles entered the solid. The circular-shaped tracks are due to alpha particles that entered straight (perpendicular to the disk), and the teardrop-shaped tracks are due to alpha particles that entered at an angle.

2. Place your disk on a microscope slide. Look through the microscope and adjust the focus until you are sure you are focused on the etched disk. Again, be sure not to mistake the microscope slide for the disk. Once you have focused on the disk, you can move the slide around to observe the tracks in different areas.

3. Count the number of tracks in your field of view and record your data on your data sheet. Move the disk to a new field of view and count and record the number of tracks in the new field of view. Repeat until you have collected and recorded data from 10 different fields of view.

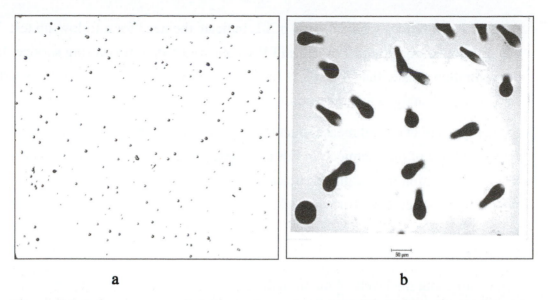

a   b

Figure 15.2 Appearance of alpha tracks on an etched CR-39 disk that had been exposed to 9 picocuries per liter for 5 days (a) at 20X magnification and (b) at high magnification, which allows you to see the different shapes of tracks.

Analyzing Evidence

1. Calculate the average number of tracks in a field of view for your disk.

2. Calculate the average number of tracks per square centimeter by dividing the average number of tracks by the area of your field of view that you calculated earlier.

3. Calculate the number of tracks per square centimeter per day of exposure (including any fraction of a day). Check with your instructor if you need help with this calculation.

4. Control plastic disks that were sent to a radon facility in which the radon level was known to be 370 picocuries per liter were found to exhibit 2370 tracks/cm^2/day. Use this relationship between picocuries per liter and tracks per square centimeter per day to calculate the radon level for *your* sample in picocuries per liter of air.

Interpreting Evidence

1. You are comparing your measurements to those in picocuries. What does pico mean? What does a curie measure? See Your Turn 6.13 in *Chemistry in Context* if needed.

2. How does your sample compare with the 4-pCi/L guidelines set by the Environmental Protection Agency?

Note: Concentrations greater than 4 pCi/L indicate the need for a follow-up test. You can do another one yourself or use an EPA-approved radon monitoring service. If a concentration greater than 4 pCi/L is found again, you should consult with experts for more detailed analysis and mitigation.

3. For many types of chemical measurement, you need to subtract a naturally occurring background level of the chemical. In this case, you don't expect to be able to measure any background radiation from alpha particles, so we don't have to subtract out a background. Why don't we expect alpha particle background radiation?

Making Claims

What can you claim about the levels of radon in different locations? What data suggest these claims? Are your claims supported by the EPA data?

Reflecting on the Investigation

1. Is your calculated radon concentration consistent with your prediction? Give some potential reasons why it is or is not as you expected.

2. Consider the following situations and determine whether radon is likely to be a problem for the occupants of the buildings. Explain why or why not.
   a. A beachside home in Mayagüez, Puerto Rico, a university town on the west coast of this volcanic Caribbean island. The home is built on a concrete slab and has screen windows.
   b. A 10-story office building in chilly Duluth, Minnesota. The building is well insulated and sealed against the winter arctic blasts.
   c. A two-story farmhouse in rural Nebraska. The basement has a dirt floor with underlying limestone rock.

3. The following health advisory was issued by the Surgeon General:
   "Indoor radon gas is a national health problem. Radon causes thousands of deaths each year. Millions of homes have elevated radon levels. Homes should be tested for radon. When elevated levels are confirmed, the problem should be corrected."
   Do an Internet search and list three things that can be done to mitigate high radon levels. Explain briefly how each of the actions you list will help solve the problem.

4. Find your hometown and college on the EPA radon map. What are the expected levels of radon in those areas? Should you or your family be concerned about radon?

Detecting Ions in Solution

Asking Questions

- What factors are important to consider when building a scientific instrument?
- How can you determine whether a solution contains ions?
- What is the relationship between ion concentration and electrical conductivity?
- Why is pure water often referred to as deionized water?

Preparing to Investigate

In this investigation, you will build a detector[1] and use it to determine whether different solutions contain electrically charged particles, called **ions** (Chapter 8 in *Chemistry in Context*). Ionic solutions contain both positively charged ions (cations) and negatively charged ions (anions). For example, when solid sodium chloride, NaCl, is dissolved in water, it breaks apart into Na^+ and Cl^- ions. These ions carry electric charges and are free to move independently, so they can conduct electricity through the solution. This property provides a very simple and useful way to test for the presence of ions: If ions are present, the solution will conduct electricity, and if no ions are present, the solution is nonconducting. Also, the magnitude of the conductance is directly proportional to the concentration of ions dissolved in the liquid.

Your detector will measure the presence of ions indirectly through determining whether the solution can conduct electricity. Specifically, the light-emitting diode (LED) in your detector will light up when the probes are immersed in a liquid that conducts electricity. This happens because conducting materials complete the circuit in your detector, and the current flow causes the LED to emit visible light.

Making Predictions

- Which of the following do you expect to conduct electricity: pure water, 1% salt solution, 1% sodium hydroxide solution, 1% HCl solution, 1% sugar solution, or 10% ethanol solution? Explain why you predict these results.

---

[1] This design for a conductivity detector was originated by F. J. Gadeck, *J. Chem. Educ.* **1987**, *64*, 628 (DOI: 10.1021/ed064p628).

- After reading *Gathering Evidence*, prepare a data sheet to house your predictions. It should also include a data table that lists the solutions and a place for you to record the conductivity measurements and observations.

Gathering Evidence

Overview of the Investigation

1. Collect the materials for building the detector.
2. Practice soldering, if necessary.
3. Assemble the detector.
4. Test the detector on a known solution.
5. Test various solutions in the laboratory.
6. Test materials outside of the laboratory.

Part I. Soldering Wires

Solder is a low-melting mixture of several metals that is melted onto wires to join them and make an electrical connection. For this investigation, you will either use an electrically heated soldering iron or "gun," or small strips of tape solder that are wrapped around the wires and then heated with a match or candle flame. The tape solder melts at a low temperature and makes a soldering iron unnecessary.

Before assembling the detector, you should practice soldering together some scrap pieces of wire as follows:

1. Using a wire stripper, strip off about 1 cm of the plastic coating from the ends of two wires.
2. Twist the bare ends of the wires together to make a mechanical connection.
3. Do ONE of the following:
    a. If you are using a soldering iron or gun, obtain a strip of solder wire. Turn on the soldering iron or gun and hold it against the end of the solder wire until the solder begins to melt. This will confirm that the soldering iron is hot. Then, touch the hot tip of the soldering iron to the wires to be joined, and hold the end of the solder wire to the heated spot. The solder wire should melt and flow smoothly over the hot wire junction. Remove the soldering iron and wait for the joint to cool. When it is cool, pull gently on the ends of the wire to be sure that the junction is formed and that the wires will not come apart.
    b. If you are using tape solder, cut a piece that is 0.5–1 cm long. Wrap the tape solder around the connected wires, and then heat the tape solder gently with a match or candle. The tape solder should melt and flow over the wires and join them together.

When the wires are cool, pull gently on the ends of the wire to be sure that the junction is formed and that the wires will not come apart.

4. Repeat steps 1–3 until you can make a simple solder joint that seems solid.

Part II. Assembling the Detector

1. Gather the parts for the conductivity detector and be sure you can correctly identify each part listed in the Figure 16.1caption. You will also need two 12-inch lengths of thin plastic tubing and a 12-inch piece of black electrical tape.

2. Prepare the film canister cap by punching four holes in it (shown). Punch the holes from the inside of the cap to avoid damaging the lip of the cap. Use a hole puncher if available. Otherwise, hold a nail in a flame with pliers until the nail is hot (not red hot) and use the nail to melt suitable holes in the top.

3. Assemble your detector as shown in Figure 16.1. Note that the letters (A through G) refer to components of the detector that are labeled in the caption. Push the two wires from the LED (F) through the two closely spaced holes in the cap of the film canister (D).

4. Locate the longer wire of the LED. If thin plastic tubing is available, cut a length so that it covers all of the wire except for about 1 cm. This tubing insulates the bare wires.

5. Twist the **long** wire of the LED and a wire from the resistor (E) together. Solder the resistor and the LED together at this joint. *Note*: If the wrong wire is attached to the resistor, the LED will be permanently damaged when current flows through the circuit.

6. If it is available, cut a length of thin tubing so that it covers all of the other resistor wire except for about 1 cm.

7. Solder the remaining resistor wire to the red wire from the battery clip (C).

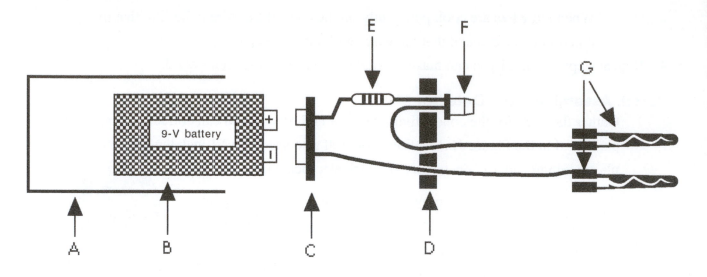

Figure 16.1. Diagram of the finished conductivity detector: (A) film canister, (B) 9-V battery, (C) battery clip, (D) canister cap with holes, (E) resistor, (F) LED, and (G) alligator clips.

8. Thread a wire through one of the remaining holes in the cap. Locate the other LED wire. If thin tubing is available, cut a length of it so that it covers all of the LED wire except for about 1 cm.

9. Solder the shorter wire of the LED to this wire.

10. Thread a second wire through the cap. Locate the black wire from the battery clip and solder it to the wire.

11. Attach alligator clips (G) to the ends of the two wires extending through the cap. First make a mechanical connection for each; then solder the connections.

12. Test your detector by connecting the battery clip to a 9-volt battery and touching the wires or the alligator clips together. Completion of the circuit should cause the LED to light up.

13. If the detector does not work, check the entire system against the diagram in Figure 16.1 to see if you have made a soldering error. Also check for loose connections. If you cannot find an error, check with your instructor.

14. If the detector works, package it in a black plastic film canister (Figure 16.1). Be careful not to pull apart any connections or have any bare wire connection touching any other bare connection.

Part III. Testing for Ions

If alligator clips are used, do not insert the metallic clips directly into the solutions. Clip them to short pieces of wire (*e.g.*, a partially unfolded paper clip) and insert that into the solutions. The wires keep the alligator clips out of the solutions and prevent them from becoming corroded. The tests can be performed in small beakers, small test tubes, or in a plastic wellplate. Be sure to record your observations as you proceed. For each test, record whether you observe a bright light, a dim light, or no light.

Note: It is important to rinse the wires with pure water between testing solutions. This will prevent contamination of your samples and ensure that you are measuring the conductivity of the current sample.

1. Test the conductivity of the following solutions by dipping the attached wires into the solution. The tips of the wires should be about half an inch below the surface of the solution and should not touch each other. If the light goes on, it means an electric current is flowing through the solution and that the solution contains ions. Record your results for these solutions.

   a. Pure water (distilled or deionized)

   b. Tap water

   c. Salt water (1% solution of sodium chloride).

2. Use the sample (pure water, tap water, or salt water) that had the brightest light to test whether different lengths of wire that are immersed in the liquid influence your conductivity results.

3. Use the same sample to determine whether the distance between the wires that are immersed in a liquid affects the sensitivity of your detector.

4. Next, test the following samples and record whether you see a bright light, a dim light, or no light from the LED. *Note*: If wire length or distance influenced the results, be sure to keep these variables constant when testing new samples.

   a. 1% sodium hydroxide (NaOH) solution

   b. 1% hydrochloric acid solution

   c. 1% sugar solution

   d. 10% ethanol solution

Part IV. Optional Activities

1. Your instructor may provide additional solutions to test in the laboratory. If so, be sure to record what the solutions are and what you observe.

2. Devise and carry out an investigation to determine how sensitive your detector is to the presence of ions in solution. Record your method and the results.

3. Now that you know how to use the conductivity tester, take it with you when you leave the lab and test at least five liquids that you encounter during the course of a day. Enter the data in your data sheet. Be prepared to report your observations. Some suggestions for liquids to try include:

    tap water from different sources

    stream or lake water

    rainwater

    sweat

    beverages such as coffee, tea, fruit juice, milk, soft drinks

    food such as a potato, a piece of apple, or other fruit

4. The detector you built can also be used to determine whether solids conduct electricity. If the probes are touched to a solid and the LED lights up, then the solid conducts electricity. Electricity is not carried through solids by ions, as in solution, but by freely moving electrons in some types of solids such as metals. With your instructor's permission, you may use your detector to test if some solid substances conduct electricity.

Analyzing Evidence

1. What was the effect of varying the length of wire immersed in the solution? Explain any differences in the conductivity measurements.

2. What was the effect of varying the distance between the wires? Describe what could have contributed to any differences in the conductivity measurements.

3. What can you conclude about the presence of ions in each of the tested solutions?

4. Did the brightness of the light vary from solution to solution? Cite specific examples. What does a dim light tell us about the number of ions in a solution?

Interpreting Evidence

1. How do your results compare with your predictions? For cases where the measured conductivity is not what you expected, how have your investigation results caused you to revise your ideas about the structures of the compounds in these solutions?

2. How would a chemist describe what is present in salt water or hydrochloric acid solutions? What investigation evidence do you have that would support this description?

Making Claims

What can you claim about the presence of ions in various solutions?

Reflecting on the Investigation

1. If you were able to test some commercial products, read the labels on the containers of these liquids and attempt to determine what substances were responsible for the liquid's conductivity. Do you see any common types of ingredients in the liquids that were conductive?

2. The purity of ultrapure water is often measured with conductivity detectors. On the basis of what you have observed, why is this a good test for water purity? In what way might it be incomplete?

3. If you wanted to make a solution of an unknown solid substance to test it for the presence of ions, would it matter whether you used pure water or tap water to make the solution? Explain.

4. What are some advantages and disadvantages of building your own lab equipment?

NOTES

Exploring Electrochemistry

Asking Questions

- How can we get electricity from chemical reactions?
- How can electricity contribute to a chemical reaction?
- What is the relationship between a galvanic cell and an electrolytic cell?
- How do electrolytes influence electrolysis?

Preparing to Investigate

Any spontaneous chemical reaction that involves the transfer of electrons from one atom or molecule to another can be used as the basis for an electrochemical cell. Consider, for example, the reaction between zinc metal and copper ions in water:

$$Zn(s) + Cu^{2+}(aq) \rightarrow Zn^{2+}(aq) + Cu(s)$$

If zinc metal, $Zn(s)$, is added to a water solution containing Cu^{2+} ions, the blue color of the copper ions disappears, the reddish color of metallic copper appears, and the zinc seems to "disappear." This process occurs through electron transfer. Each zinc atom loses two electrons and forms a +2 ion. Each copper ion adds two electrons and forms metallic, neutral copper. The zinc ions are colorless and soluble in water, and the copper precipitates out. As explained in Chapters 1 and 7 in *Chemistry in Context*, we can break this up into two half-reactions:

$$Zn(s) \rightarrow Zn^{2+}(aq) + 2\ e^-$$
$$Cu^{2+}(aq) + 2\ e^- \rightarrow Cu(s)$$

To get useful electric energy from this reaction, the two half-reactions have to occur in two separate locations, but they must be connected so that the electrons can flow through an external wire. The wire can be connected to something (such as a motor, light bulb, or meter) to show that an electrical current is produced. A device with these components is called a **galvanic cell**, although the nonscientific and slightly incorrect name **battery** is often used. In Part I of this investigation, you will investigate how to assemble several working galvanic cells and use a voltmeter to measure the voltages produced.

Reflecting on the Investigation

1. Imagine that you are marooned on an isolated island with a few other people. After doing this investigation, what could you offer to the group that would be useful for supplying energy? Suggest something specific you could do with a few pieces of metal and food scraps or some nearby salt deposits.

2. In Part IIB, during the electrolysis of the KI solution, either H_2O or KI could react at each of the electrodes. Thus, although only one reaction will actually occur at each electrode, we must consider two possible reactions for the red electrode and two possible reactions for the black electrode. The half-reactions that could occur are the following:

- At the electrode where electrons are consumed:

    Either: $4\ H_2O(l) + 4\ e^- \rightarrow 2\ H_2(g) + 4\ OH^-\ (aq)$ (gas and basic solution formed)

    Or: $K^+(aq) + e^- \rightarrow K(s)$ (metallic potassium formed)

- At the electrode where electrodes are released:

    Either: $2\ H_2O(l) \rightarrow O_2(g) + 4\ e^- + 4\ H^+(aq)$ (gas and acidic solution formed)

    Or: $2\ I^-\ (aq) \rightarrow I_2(aq) + 2\ e^-$ (yellow-brown elemental I_2 formed)

    a. From your observations, what formed at the black-wire lead? Explain your reasoning.

    b. From your observations, what formed at the red-wire lead? Explain your reasoning.

    c. Compare your electrolysis of water to that of the KI solution. Which product from the water electrolysis was also produced in the KI electrolysis?

    d. Which product from the water electrolysis was NOT produced in the KI electrolysis? What was formed instead?

    e. Electrolysis will always produce the products that are easiest to make from the starting materials. What can you conclude about how easy it is to remove electrons from iodine in KI versus how easy it is to remove electrons from oxygen in H_2O?

Analyzing Water

Asking Questions

- What compounds and ions, besides water, are often found in water?
- How does human activity affect water purity?
- Describe some environmental contamination that can affect the pH of water.
- What ions make water "hard"? What problems can result from having hard water?
- Look at a periodic table and identify the locations of magnesium and calcium. From the discussion of atomic structure and periodicity in Chapter 1 of *Chemistry in Context,* explain why these two elements may react in a similar manner to each other.
- Why does chloride ion concentration in water increase with human activity and pollution?
- How do chloride ions differ from atoms of chlorine? Why might elemental chlorine be added to water?

Preparing to Investigate

What we call "water" is rarely, if ever, pure H_2O. When you pour yourself a glass of water from the tap or from a bottle, that water contains dissolved compounds and ions that, for better or worse, affect the purity and flavor of the water. Additionally, environmental factors such as local geology and atmospheric pollution can affect the purity of rainwater, lakes, and rivers. As discussed in Chapter 8 of *Chemistry in Context* the actual composition of water is affected by geology and human activities. In this experiment, you will use various titration procedures to analyze ions present in water. You will collect or be provided with water samples, and design an experiment to investigate the composition of the samples.

The procedures in this investigation will allow you to test for the following characteristics of your water samples:

1. **pH**

The pH of pure water is 7, but water can become slightly acidic or basic due to a number of factors. For example, rain can become acidic (lower the pH) by reacting with atmospheric SO_x and dissolved CO_2. The presence of bicarbonates, in contrast, will make water become more basic (raise the pH of water). To measure the pH, you will use the procedure described in the *Laboratory Methods* section of this book. If you have not yet used a pH meter, your instructor will demonstrate how to use it. It is important to learn the proper use of this instrument so that you do not damage it. Alternatively, you can use pH test strips to measure the pH of your samples. This is faster, but less precise, than using a pH meter.

2. Concentration of Bicarbonate

The bicarbonate (HCO_3^-) concentration in water is most often affected by the presence of carbonaceous rocks, such as limestone, in the local geology. You will perform a titration analysis of bicarbonate. If you haven't yet performed a titration, you should review the procedure in the *Laboratory Methods* section of this book.

3. Water Hardness

The combined concentration of calcium ions (Ca^{2+}) and magnesium ions (Mg^{2+}) is called the "total hardness" of water. A high concentration of dissolved minerals that contain these alkaline earth ions can lead to uncomfortable, but not dangerous, consequences in household settings, such as the build-up of soap scum and formation of spots or a scaly film (calcium carbonate) on clean dishes. You will analyze the combined concentration of calcium and magnesium using a titration analysis.

4. Chloride Ion

The chloride ion (Cl^-) content of natural, unpolluted surface water depends in large part on the geology of the area. In places where the surface water normally has very little chloride, a higher chloride concentration implies a source from human activity, such as manufacturing or sewage discharge. Measuring the chloride concentration from water samples taken from streams or rivers in your local area will indicate the degree to which they are impacted by human activities close to waterways.

Note that ionic chloride (Cl^-) is quite different than elemental chlorine (Cl_2), which is a gas at room temperature. Chlorine is added as a disinfectant to drinking water in small amounts (less than 1 ppm) and to swimming pools in somewhat larger amounts (about 5 ppm). Chlorine will produce some chloride as it reacts, but in most cases it is not a major source of chloride. You will measure the chloride content in water by performing a titration.

5. Concentration of Total Dissolved Solids

The total dissolved solids (TDSs) is a way to measure the amount of dissolved material. TDSs consist mostly of ionic substances, or salts, since water is an excellent solvent for ionic compounds. You will measure TDSs by heating the sample to evaporate the water, and you will then measure the mass of any solid materials that remain. If you analyze samples of bottled water, the label may indicate the TDSs concentration, so you can easily compare your measurement.

Making Predictions

In this investigation you will get to decide, along with your classmates and with input from your instructor, what questions to ask and how to design an experiment to answer them. On the basis of what you have learned about water, you should ask some questions that you can answer using the methods outlined in this investigation. Some example questions might be:

- How does the pH of rain or snow differ in various locations?

- How does water purity change upstream and downstream of my town?

- How does the water hardness in my school compare to that in my house or in my classmates' hometowns?

- How do the concentrations of different ions in tap water compare to bottled water?

- How does distilling water or running it through an ion-exchange column change the concentration of ions?

You will be given collection vessels and instructions on how to properly collect water samples. Alternatively, you may be given one or more water samples to analyze by your instructor. You can combine your analytical data with that of your classmates to collectively answer one or more questions that you pose.

Gathering Evidence

Overview of the Investigation
1. Collect and label the water samples that you will analyze to answer your posed questions.
2. Determine which analyses are important to answer your questions.
3. Conduct the analyses.
4. Assemble the analytical information to compare results for different samples.

Part I. Measuring pH

Review the instructions in the *Laboratory Methods* section for measuring pH using a pH meter or pH test strips. Follow any directions given by your instructor for the operation of the instruments.

Part II. Bicarbonate Analysis

Analysis of bicarbonate involves a titration analysis, and is based on the fact that bicarbonate is a base that can react with an acid such as hydrochloric acid (HCl) as shown here.

$$HCO_3^-(aq) + HCl(aq) \rightarrow H_2O(l) + CO_2(g) + Cl^-(aq)$$

A colored indicator, methyl orange, will change color from yellow to orange when the pH drops below 4. This occurs when an excess of HCl is present and indicates that all of the bicarbonate ion has reacted.

1. Gather the materials you need for your titration, including a plastic wellplate, small stirring rod, and two pipets. Place the wellplate on a white sheet of paper so that the color change can be easily seen, and label your pipets "water sample" and "HCl." Labeling is crucial because it is easy to get the pipets mixed up! If necessary, practice using the pipet so that you can reliably dispense one full drop of liquid at a time into the wellplate. A pipet practice procedure is given in *Laboratory Methods*.

2. Obtain about 10 mL of HCl solution in a clean, dry beaker, and record its concentration.

3. Add 10 drops of your water sample into a well of the wellplate (record the well position), and then add 1 drop of methyl orange indicator. Your solution should be yellow.

4. Add the HCl solution one drop at a time, counting the drops and stirring gently after each addition. Carefully observe what happens. Watch for the first color change from yellow to orange that persists after the mixture is stirred. Record the number of drops used.

5. To aid in recognizing the color change for subsequent titrations, add a few more drops of your water sample to the well. The color should go back to yellow. Save this for color comparison. In your other titrations, you will be looking for the first persistent color change away from this yellow.

6. From the result of the first titration, decide how much water to use for subsequent titrations so as to require a convenient amount of HCl. If possible, you should use at least 10 drops but no more than 30 drops of HCl.

7. Carefully add the desired number of drops of water to three more wells in your wellplate.

8. Add 1–2 drops of methyl orange indicator to each well with water.

9. Titrate each of the three samples, one at a time, using the HCl solution. Record the number of drops used. If the results do not show adequate consistency, do one or two additional titrations.

10. Average your results and calculate the amount of bicarbonate present in the sample using the equation below.

$$\text{molarity of HCO}_3^- = -\text{molarity of HCl} \times \frac{\text{number of drops of HCl solution}}{\text{number of drops of water sample}}$$

To convert molarity to ppm, multiply your calculated molarity by the molar mass of bicarbonate, 61,000 mg/mol.

Part III. Water Hardness

Your analysis of water hardness will be based on the chemical reaction of Ca^{2+} and Mg^{2+} with an ion called ethylenediaminetetraacetate (EDTA), which has the molecular formula $C_{10}H_{12}N_2O_8{}^{4-}$. In the chemical reaction you will perform, the EDTA ion has two protons (H^+) attached, which results in a -2 charge ($-4 + +2 = -2$). The ion with the extra protons is written H_2EDTA^{2-}. The reaction you will carry out involves one calcium or magnesium ion reacting with one H_2EDTA^{2-} ion, as shown below.

$$Ca^{2+} + H_2EDTA^{2-} \rightarrow Ca(EDTA)^{2-} + 2\,H^+$$

To do the analysis, you will titrate a measured amount of water sample with an EDTA solution of known concentration until it completely reacts with the calcium and magnesium ions in the sample. To see the reaction, you will use a small amount of an indicator called calmagite. At pH 10 (an alkaline solution), the indicator changes to a deep blue color, but in the presence of metal ions it is red. Thus, if a drop of indicator is added to a solution containing calcium and magnesium ions, the solution will have a reddish color. When enough EDTA has been added to react with all of the calcium and magnesium, the solution will turn purple and then blue.

Obtain the water samples to be analyzed. Note that some water samples, especially tap water in some buildings, may not change color to pure blue due to traces of iron or copper in the water. If this is the case with your sample, check with your instructor before proceeding. It may be necessary to titrate to a purple color rather than blue.

1. To test the color of your indicator solution, add 10 drops of pure water to a well of the wellplate that has been placed on a white surface, and then add 2 drops of pH 10 buffer and 1 drop of calmagite indicator. Stir the mixture with a small stirring rod and note the color. If it is not blue, add 1 drop of EDTA solution to produce a bright clear blue color with no trace of red or purple. This is the color to look for at the end of each titration. Leave this solution in your wellplate as a reference for when you are doing your titrations.

2. Dispense 10 drops of your water sample into a clean well in the wellplate, and record the well position.

3. Add two drops of pH 10 buffer solution, and then 1 drop of calmagite indicator solution, in that order. Then, add the EDTA solution slowly, one drop at a time, counting drops and stirring after each addition. Continue until the color starts to change from red to purple; the reaction is slow so wait a few moments before adding another drop, stirring patiently until the color becomes pure blue. Count the drops of EDTA added and enter the final number into your data table.

4. Repeat the procedure at least twice more for each water sample, changing the initial number of drops of water sample if necessary.

5. Calculate the water hardness using the following equation.

$$\text{molarity of Ca}^{2+} + \text{Mg}^{2+} = \text{molarity of EDTA} \times \frac{\text{number of drops of EDTA solution}}{\text{number of drops of water sample}}$$

To convert the calculated molarity into ppm, you need to multiply the molarity by the molar mass of the ion. This analysis measures the *combined* concentration of the calcium and magnesium ions, but since calcium is almost always present in much higher concentration than magnesium we will simplify the calculation by assuming that all of the measured hardness is due to calcium. You can simply multiply the molarity calculated by the molar mass of calcium, 40,000 mg/mol.

Part IV. Chloride Ion Concentration

You will measure the chloride content in water by performing a titration in which the chloride reacts with silver nitrate ($AgNO_3$) to form an insoluble white compound, silver chloride (AgCl). This is called the Mohr Method of analysis. Other ions present in the water do not participate in this reaction.

$$AgNO_3(aq) + Cl^-(aq) \rightarrow AgCl(s) + NO_3^-(aq)$$

To know when enough silver nitrate has been added to react with all of the chloride in your water sample, you will use a solution of sodium chromate (Na_2CrO_4) as an indicator. Chromate ions are yellow, but react with silver ions to form silver chromate, a red solid.

$$2\,Ag^+(aq) + CrO_4^{2-}(aq) \rightarrow Ag_2CrO_4(s)$$

Thus, as silver nitrate is added to your water sample, the chloride precipitates as white silver chloride. After all of the chloride has been removed, additional silver ion will react with the chromate to form a red insoluble precipitate of silver chromate. The appearance of this red precipitate signals the endpoint of the titration.

STOP! Sodium chromate, the indicator used in this investigation, is a carcinogen. Define the term *carcinogen*. What precautions should you take when working with this compound?

1. Gather the materials you need for your titration, including a plastic wellplate, small stirring rod, and two pipets. Place the wellplate on a white sheet of paper so that the color change can be easily seen, and label your pipets "water sample" and "silver nitrate." Labeling is crucial because it is easy to get the pipets mixed up!

2. Dispense about 10 mL of silver nitrate into a clean *dry* beaker or other container. Label the container and record the exact concentration of the solution.

CAUTION! Silver nitrate will stain your skin black, so be careful not to get it on your hands. The stain may develop the next day, and it is harmless and will wear off in a few days.

3. Perform a practice titration using tap water to see the color change:

   a. Use your "water sample" pipet to add 10 drops of tap water to a well in the wellplate.

   b. Add 1 drop of sodium chromate indicator and stir. The mixture should be pale yellow-green.

CAUTION! Sodium chromate is a carcinogen! Avoid contact with your skin.

   c. Use your other pipet to add $AgNO_3$ solution, one drop at a time with gentle stirring. Count the drops and carefully observe what happens. As each drop is added, you will observe a cloudy appearance as solid silver nitrate is formed. A reddish color may appear and then disappear on stirring. As you near the endpoint, the mixture in the well will have an orange tint, and the first appearance of a uniform tint that does not disappear with stirring signals the endpoint. Make a note of how many drops were used. Then, add 1 or 2 more drops of silver nitrate with stirring. The color will become darker orange or red due to formation of more Ag_2CrO_4.

d. Finally, add a few more drops of tap water, with stirring, until the color reverts back to a milky yellow-green. Save this mixture for color comparison with your water samples.

4. You should then analyze your water samples. The goal of your titration is to catch the point where one drop of $AgNO_3$ produces the first persistent color change. It should not be a dramatic color change, but rather a slight orange tint. This is the endpoint where all of the chloride has been used up by reacting with $AgNO_3$. This color change may be difficult to recognize. If so, use your practice titration mixture for comparison. You are looking for the first change away from that color.

a. For each water sample, dispense 20 drops of the sample into a well. Add 1 drop of indicator, and then add $AgNO_3$, 1 drop at a time, with stirring, as described above. Depending on the nature of the sample, it may have very little chloride, requiring only a few drops of silver nitrate; or the chloride concentration may be very high, requiring a large quantity of silver nitrate. Record the drops of silver nitrate used.

b. Repeat the measurement at least three more times. Depending on the results from this first titration, you may decide to use more or less than 20 drops of your water sample. Dispense samples of the desired size into three more wells, and record all of your data.

c. If you miss the endpoint, lose count of drops, or are otherwise uncertain of a result for any reason, it is good scientific practice to *not* omit the data. Make a note on your data table about what you think went wrong, draw a single line through the incorrect data, and do another titration. This is good scientific practice.

d. You may repeat the procedure for additional water samples. Be sure to rinse your pipet and beaker used for your water samples between each measurement, first with tap water and then with distilled or deionized water and lastly with the new sample to be analyzed.

5. After performing your titrations you should have at least four results for each water sample. Use the following equation to calculate the molarity of chloride in your water sample.

$$\text{molarity of Cl}^- = \text{molarity of AgNO}_3 \times \frac{\text{number of drops of AgNO}_3 \text{ solution}}{\text{number of drops of water sample}}$$

You can convert the molarity to ppm by multiplying your result by 35,500 mg/mol, the molar mass of chloride.

Part V. Total Dissolved Solids (TDSs)

You will measure TDSs by heating the sample to evaporate the water and determine the mass of any solid materials that remain.

1. Check that a hot plate has been turned on and is ready to use. Then, label two clean, dry 50-mL beakers with your sample and your initials. Weigh each beaker and record the mass to the nearest milligram. The use of an analytical balance is described in detail in the *Laboratory Methods* section.

2. Use a graduated cylinder to measure 20 mL of your water sample and add it to one of your weighed beakers. Repeat this measurement for the other beaker.

3. Place the beakers on the hot plate. Adjust the temperature so that the water gently boils. Continue heating until the water has *completely* boiled off from the solutions.

4. Use tongs or an oven glove to remove the beakers from the hot plate. Allow them to cool completely. Carefully inspect the contents to ensure that the beakers are completely dry and that there is not any moisture remaining on the bottom or sides. If you see moisture, place the beaker back on the hot plate until it is completely dry.

5. Weigh the beakers and record the mass. Subtract the initial mass from the final mass to calculate the mass of recovered solid, and convert it to from g to mg if necessary (1000 mg = 1 g). Convert the amount of water used into liters (1000 mL = 1 L). Then, use the equation below to calculate the total dissolved solids in ppm.

$$\text{TDS in ppm} = \frac{\text{mass of solid recovered (in mg)}}{\text{volume of water sample (in L)}}$$

Clean up

When you are finished with your analyses, follow all instructions regarding waste disposal. Dump the contents of the wellplates into designated waste containers, and rinse and drain them thoroughly. Your instructor will tell you whether to wash or dispose of the plastic pipets. The remaining glass and plastic items should be rinsed thoroughly with deionized or distilled water and left upside-down to drain.

Analyzing Evidence

1. With your lab partners, complete the necessary calculations for all analyses that you performed. Include units on your final values. Remember that aside from pH, all other measurements should be in units of ppm.

2. Look carefully at the data from each replicate set of titration measurements. Are the results consistent? Do you notice any results that seem out of line with the others? If so, it is likely that a mistake was made in that titration, and it is legitimate to omit that result. If you decide to omit any outlier results from your analysis, simply cross out the result with a single line and write a note about what you think went wrong.

3. Calculate the average for each replicate set of titrations, leaving out the data you have decided to omit.

4. Share your results with others from your class by writing them on the board or following another procedure specified by your instructor. Record your classmates' results.

Interpreting Evidence

1. What is the advantage of doing three or more trials of each titration, especially since you are doing exactly the same thing each time?

2. What patterns do you see in the assembled data from your class? Compare pH, bicarbonate, total hardness, chloride concentration, and TDSs from the samples you and your classmates measured.

3. It is important in scientific investigations to quantify the errors that may be present in your measurements. In these titrations, there may be an uncertainty of at least one drop in identifying the exact endpoint. If you use 20 drops of your analysis solution in a titration, what is the percent uncertainty in your measurement contributed by this one extra drop?

$$\% \text{ error} = \frac{\text{number of drops you are in error}}{\text{number of drops used}} \times 100\%$$

Making Claims

Use the data collected by your class to address the questions you asked in the *Making Predictions* section. What can you claim about how your water samples were affected by geology, weather, or human activity?

Reflecting on the Investigation

1. The pH of water for human consumption should be close to neutral. Did you measure any samples that would be unfit for consumption by this criterion?

2. Do the differences you observed in ion concentrations make sense based on what you know about the sources of these ions in water?

3. List some of the major non-industrial sources of chloride ion in natural waters. Which ones do you think are important in your area?

4. If you measured the ion concentrations or TDSs of bottled water, how closely do your results match the numbers listed on the label?

5. Do you feel that drinking bottled water is more, less, or equally safe than drinking tap water? Write a paragraph explaining your position.

6. On the basis of what you have learned in this investigation, do you consider chloride to be a serious water pollutant? Why or why not?

NOTES

Reactions of Acids with Common Substances

Asking Questions

- What is an acid?

- What sorts of elements or compounds react with acids?

- How and why does acid rain affect buildings and living organisms?

Preparing to Investigate

Acids are substances that contain hydrogen ions (H^+) or react with water to form hydrogen ions. An important category of chemical substances, acids are described in detail in Chapter 8 of *Chemistry in Context*. They are found in a variety of things such as foods, our bodies, and rain or snow. Acids undergo chemical reactions with many substances. Some of these reactions are desirable, while others can be damaging. For instance, the acids found in rain or snow can have destructive effects on building materials, and other adverse environmental consequences.

In this investigation, you will observe the reactions of three common acids: hydrochloric acid (HCl), sulfuric acid (H_2SO_4), and nitric acid (HNO_3). While all of these compounds are acids, you will see that not all acids react in the same way. Careful observation will allow you to conclude that the rates of reaction of acids with materials vary with concentration. To make observations in a limited amount of time, you will use acid solutions that are more concentrated than those found in rain or snow. Still, the same reactions do occur with acid rain but on a longer time scale.

You will study the reactions of these acids with some familiar substances. You will observe reactions with four common metals—zinc, copper, nickel, and aluminum. Most metals react with acids to produce colorless hydrogen gas (H_2) and ionic compounds. You will also see the effects of acid on marble, a form of calcium carbonate ($CaCO_3$). Building materials vary in their reactivity with acid; some, like marble, readily react with acids, while others, such as granite, do not visibly react. Lastly, you will see the reaction of acid with egg white, a protein that represents living matter. The reaction of proteins with acid is unlike that of the other materials. In aqueous solutions, protein molecules have a preferred three-dimensional shape that is pH-dependent. Addition of an acid lowers the pH and changes the protein shape so that the protein becomes less soluble and the formation of a solid may be visible. In addition, some acids will chemically react with the protein molecule itself.

Concentrations of the acids in this investigation are expressed as **molarity.** The molarity of a substance in solution is defined as the number of moles of that substance in 1 liter of solution. It is expressed in units of mol/L or M.

Making Predictions

- Locate zinc, copper, nickel, and aluminum on the periodic table. Do you expect their reactions with acid to be similar to each other or different? Why?

- Do you expect to see greater reaction from 6-M or 0.6-M sulfuric acid? Why?

- After reading *Gathering Evidence*, prepare a table to record your observations from the investigation of the four different acid solutions with the six different samples.

Gathering Evidence

Overview of the Investigation

1. Test the reactions of four acid solutions with marble chips (calcium carbonate).

2. Test the acid solutions with four different metals.

3. Expose egg whites to the acid solutions.

 STOP! You should always wear eye protection in the laboratory, but this is especially important when working with acids. Safety glasses are absolutely essential to protect your eyes from any splashes.

Part I. Preparing the Metal Samples

The tests with metals will work best when the metal surface has been freshly cleaned to remove any corrosion or coating. Therefore, you should obtain the following sets of metal pieces and clean at least one surface of each piece with sandpaper, emery paper, or steel wool.

- Zinc: These samples can be either thin strips of zinc or galvanized nails, which are made of iron coated with zinc. When clean, the surface should be bright and shiny.

- Copper: These samples may be short pieces of heavy-gauge copper wire or small pellets of copper shot. The surface should be bright and shiny.

- Nickel: Fresh, shiny paper clips are usually made of iron or steel with a coating of nickel. If the paper clips are new and shiny, they shouldn't need any cleaning.

- Aluminum: These samples can be made by cutting small strips out of an empty beverage can and then cleaning one surface to expose fresh metal.

Part II. Preparing the Wellplate

You will use all 24 wells of your wellplate for this investigation, and it is important to keep track of which acid is in each well. You will fill the wells with the acids, and then drop your test objects into each well to observe the reaction.

CAUTION! Be careful not to spill acid on your skin, clothing, books, or papers. If you do spill any acid on your skin, wash it off promptly with large amounts of water. Continue to run water on the affected area for several minutes. Notify your instructor immediately in the event of a spill on any person or workspace.

1. Place your clean, dry wellplate on a white sheet of paper. Label the rows and columns on the paper according to the diagram in Figure 19.1.

2. Add one dropper (about 20–30 drops, or 1–1.5 mL) of 6 M hydrochloric acid (HCl) to each of the wells in Row A. The wells should be filled less than half way.

3. Similarly, add one dropper of 6-M sulfuric acid (H_2SO_4) to each well in Row B.

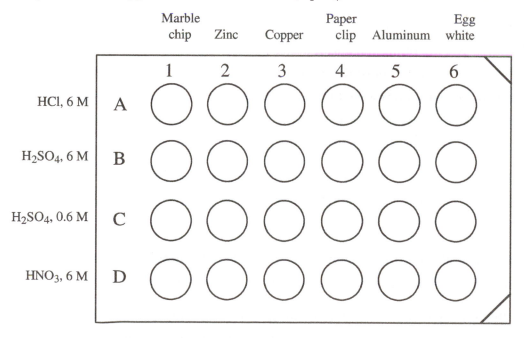

Figure 19.1. Diagram of the wellplate setup.

Making Claims

What can you claim about the reactions of acids with different substances? Include specific examples from your investigation data.

Reflecting on the Investigation

1. The reactivity of marble with sulfuric acid seems curiously out of line with all of the other observations. If you observed carefully, you may have seen a brief burst of gas bubbles, which quickly stopped. This is because one product of the reaction, calcium sulfate ($CaSO_4$), is insoluble and provides a protective coating on the marble, which slows the reaction dramatically. Does this mean that acid rain containing sulfuric acid should have little or no effect on statues and buildings made of marble? Explain your answer.

2. Iron reacts readily with acid, but is also an important structural material for buildings. Based on your observations, how effective are zinc or nickel coatings as a protection for iron against acid rain?

3. Copper has been important since ancient times as a building material used in roofing structures. On the basis of your observations, is copper a good material for this purpose? *Note*: You may have observed a green color on the roofs of public buildings or monuments such as the Statue of Liberty. This is a form of copper carbonate, formed by slow reaction of copper with nitric acid plus carbon dioxide from the atmosphere.

4. What differences did you observe for reactions of aluminum compared to the other metals? Aluminum easily forms an unreactive oxide coating. How does this account for the observed differences in aluminum's reactivity?

5. Stomach acid is approximately 0.16-M HCl. Based on your observations of the reaction of acids with egg white, what affects might stomach acid have on protein you eat? Would you expect the reaction of stomach acid with protein or food to be faster or slower than the reaction you observed between egg white and 6-M HCl?

Characterizing Acidic and Basic Materials

Asking Questions

- What common substances are acidic or basic?

- How can acidity and basicity be measured?

- What are the benefits and challenges of measuring pH with a calibrated pH meter?

Preparing to Investigate

The acidity or basicity of substances can be determined by measuring their **pH**. The pH scale is a convenient way of expressing hydrogen ion concentration in solutions. Chapter 8 in *Chemistry in Context* discusses acidity and some of the consequences of acidification of rain and ocean water. pH is defined as the negative of the logarithm of the hydrogen ion concentration, $[H^+]$, in **molarity** (mol/L). *Note:* Refer to Appendix 3 in *Chemistry in Context* for more information about the math involved with logarithms.

$$pH = -\log [H^+]$$

Modern pH meters are easy to use and generally reliable, but their sensitivity means that they are highly susceptible to interference. If you use an electronic pH meter to record your pH values, you should carefully follow the directions for using it so you record accurate values without damaging the instrument. Alternatively, your instructor may have you measure pH using paper test strips that have been treated with dyes, which change color depending on pH. This method is less accurate, but can be used.

Making Predictions

After reading *Gathering Evidence*, prepare a data table to record your predictions and results. Predict whether the substances you will test are acidic (pH < 7) or basic (pH > 7).

Gathering Evidence

Overview of the Investigation

1. Learn to operate a pH meter and calibrate the meter.

2. Measure the pH of pure acid and base solutions.

3. Measure the pH of various foods and household products.

4. Measure the pH of tap water, rainwater, and surface water.

5. Measure the pH of water containing atmospheric gases CO_2 (and optional SO_2).

Part I. Operating and Calibrating the pH Meter

Detailed instructions on using a pH meter can be found in the *Laboratory Methods* section of this laboratory manual, and your instructor may provide additional guidelines for the use and care of your specific pH meter. Calibrate the pH meter using reference buffer solutions with pH 7 and pH 4. Before each measurement you should wash the electrode with pure water and gently blot it dry with a soft tissue. The electrode should always remain submerged in water or another liquid to ensure that it never dries out. The electrodes must be handled with care because they are both fragile and expensive. When measuring pH, you should start with the *least* acidic solution and proceed to the *most* acidic solution.

Part II. Measuring the pH of Chemical Solutions of Known Concentration

1. Calculate the pH of a 0.0001 M solution of HCl. Then, measure the pH of the solution.

2. Repeat Step 1 for solutions of 0.001 M and 0.01 M HCl, and for 0.001 M NaOH. Be sure to calculate your pH *before* doing the measurement.

Part III. Measuring the pH of Foods and Household Substances

Now that you know how to measure pH, you should measure the pH of some common substances. Samples may be provided by the instructor or brought in by students. In each case, put a <u>small</u> amount of the substance in one of the wells of a plastic wellplate, or in a small test tube or beaker. Use only enough so that the bulb end of the electrode is immersed. Check with your instructor if you have questions. Remember to rinse the electrode thoroughly and blot it dry before each measurement.

Below are some ideas for substances to test. You should measure the pH of six to eight substances.

- vinegar
- lemon juice
- fruit juices
- soft drinks
- coffee or tea

<u>The following substances must be diluted before testing the pH. To do this, mix a few drops of liquid or a very small scoop of solid with pure water.</u>

- liquid dish detergent
- dishwasher detergent
- laundry detergent
- shampoo or hand soap
- household ammonia
- liquid laundry bleach
- baking soda (sodium bicarbonate)
- drain cleaner (*see **Caution!** note below*)

CAUTION! Drain cleaners can be extremely caustic. They are designed to dissolve hair and other debris. Be especially cautious about getting any drain cleaner on your skin, and always wear your safety glasses. In case of any skin contact, wash with copious amounts of water and notify the instructor immediately.

Part IV. Measuring the pH of Tap Water, Rainwater, and Surface Water

1. Measure the pH of tap water from the laboratory sink, a drinking fountain, or elsewhere on campus.

2. The pH of rain or snow is particularly interesting because of concern about "acid rain." Your instructor may provide samples that have been collected recently or you can collect your own sample using the procedure described in Investigation 21. Remember that rain is nearly pure water and any acid present will be dilute, which can make pH changes difficult to measure.

3. It is instructive to compare the pH of rain to the pH of river water. If river samples are available in the lab, check the pH.

4. If water from a swimming pool is available, check its pH.

Part V. Measuring the pH of Water Containing CO_2 (or SO_2)

1. Test the pH of water that is saturated with carbon dioxide. You can use a sample of seltzer water that has been allowed to go "flat," or you can generate carbon dioxide as described in Investigation 1. Test the pH of the water containing CO_2.

2. Test the pH of water that is saturated with air. Remember that CO_2 is only a very small fraction of the gases that make up air.

3. Test the pH of water that is saturated with SO_2.

CAUTION! Sulfur dioxide is a toxic gas, and some individuals are allergic to SO_2. This test should be done *only* in a fume hood and should not be attempted by individuals who are allergic to SO_2.

   a. Make SO_2 in a zipper bag. Fill a plastic pipet with 6-M sulfuric acid and place it in a quart-size zipper bag. Add 1–2 grams of sodium sulfite (Na_2SO_3) to the bag. Close the bag securely, excluding the air, and then squeeze the pipet so that the chemicals will mix and react.

b. Saturate water with SO_2. With the bag lying on the lab bench, <u>very cautiously</u> open a corner of the bag, squeeze the air out of a dry plastic pipet, insert it into the bag, and fill the pipet with SO_2. <u>Reseal the bag immediately</u>. Slowly bubble this gas into pure water in a test tube or wellplate, and then test the pH of the water.

c. Dispose of your SO_2 sample into a designated waste container in the fume hood.

Clean up

Dispose of your samples in appropriate waste containers as instructed. Clean your pH electrode and store it in water or another solution as designated by your instructor.

Analyzing Evidence

Look carefully at your results table.

1. In Part II, did your calculated pH values match your measured pH values?

2. In Parts III and IV, did you correctly predict which substances were acidic and which were basic? How well did your predictions match up with your pH measurements?

3. In Part V, were your solutions acidic or basic?

Interpreting Evidence

1. Can you propose a general rule for how the pH should change (up or down and by how much) when the acid concentration increases by a factor of 10? For example, when HCl concentrations change from 0.0001 M to 0.001 M to 0.01 M, the acid concentrations increase by a factor of 10 each time.

2. Are tap water and surface water acidic or basic? What substances may account for this?

3. Is the pH of water from a river different from the pH of collected rain? If so, suggest an explanation.

4. Is the pH of swimming pool water different than tap water? If so, suggest an explanation.

5. Were the measured pH values of CO_2 and SO_2 in water what you expected? Write chemical equations for the reaction of each gas with water.

Making Claims

What can you claim about the pH of different substances?

Reflecting on the Investigation

1. Explain why pH is a useful way of describing acid and base solutions over a very wide range of concentrations.

2. Can you draw any conclusions about the pH of different categories of common substances? What is the general pH of beverages? Of soaps and detergents? Of substances that normally touch human skin?

3. In Part IV, was the rain acidic or basic? Since rain is formed by evaporation and condensation in clouds, why is the pH not the same as pure water?

4. In your own words, describe how (a) increased concentrations of SO_2 lead to acid rain; or (b) increased concentrations of CO_2 lead to ocean acidification.

NOTES

Acid Rain

Asking Questions

- What is the normal pH of rain?

- Does acid rain fall where you live?

- What substances in rain are responsible for making it acidic?

- What are some of the consequences to infrastructure and ecosystems of acid rain?

Preparing to Investigate

The **pH** scale is a convenient way of expressing the hydrogen ion concentration in solutions. pH is defined as the negative of the logarithm of the hydrogen ion **molarity**, $[H^+]$. *Note:* Refer to Appendix 3 in *Chemistry in Context* for more information about the math involved with logarithms.

$$pH = -\log[H^+]$$

All water samples contain some hydrogen ions. **Acidic** solutions have higher concentrations of hydrogen ions than do **basic** solutions. In water at 25 °C, the product of the molarities of H^+ and OH^- is always 1.0×10^{-14}. In pure water, the concentrations of H^+ and OH^- are equal to each other, both 1.0×10^{-7} M, so the pH is 7. In acidic solutions, the H^+ ion is present in excess compared to OH^-; the concentration will be higher than 1.0×10^{-7} M, and thus the pH will be less than 7. Chapter 8 in *Chemistry in Context* discusses pH in greater detail.

Pure rainwater has a pH of approximately 5.6, due to the presence of carbon dioxide. CO_2 reacts with water to produce carbonic acid, which then dissociates into hydrogen ions (H^+) and bicarbonate ions (HCO_3^-).

$$CO_2(g) + H_2O \rightarrow H_2CO_3 \rightarrow H^+ + HCO_3^-$$

When rain has a pH below 5.6, this means that other acids must be present. Rain can be classified as follows:

Rain classification	pH
Pure rain	5.6
Slightly acidic	5.0-5.6
Moderately acidic	4.5-5.0
Highly acidic	4.0-4.5
Extremely acidic	Below 4.0

The average pH of rain varies around the world. To learn more, you may wish to visit the U.S. Environmental Protection Agency website and view acid rain maps that are available there. If you live outside the U.S., you can do a web search for similar maps of your country.

Making Predictions

After reading *Gathering Evidence*, prepare a data table to record your predictions and pH measurements. Predict the relative acidity of your rain samples. How do you think acidity will vary based on where and when they were collected?

Gathering Evidence

Overview of the Investigation

1. Prepare sample containers for collecting rain samples.
2. Select sites for the collection containers and collect the samples.
3. Prepare the samples for pH measurement.
4. Calibrate the pH meter and measure the pH of the rain samples.

Part I. Preparing Sample Containers

Samples should be collected in plastic containers, not glass, and it is important that the containers be thoroughly cleaned in advance. The recommended cleaning procedure is to first rinse each container with 6-M hydrochloric acid, then five times with tap water and finally five rinses with distilled or deionized water. Once cleaned and dried, the containers should be capped or covered to keep them clean.

CAUTION! 6-M HCl is a corrosive liquid. Wear gloves and a lab apron or coat along with your safety glasses. If any acid contacts your skin, make sure to rinse with plenty of water and inform your instructor.

Part II. Collecting, Handling, and Storing Rain Samples

Your instructor may assign you a location for collecting rainwater samples, or your class may discuss and decide collectively on the best collection locations for answering the questions you have. If your bottles have a large opening, they can be set out to collect rain. Alternatively, to obtain larger samples, you may need to use a funnel, which should also be cleaned in advance. With a funnel, the bottle may need to be supported in an upright position by, for example, placing it inside a metal can that is nailed to a post. The collection container should be placed in an unobstructed location where it will not be disturbed and ideally will be several feet above ground level. This will minimize contamination of the sample by dirt and dust.

A single sample can be collected for an entire rain event. However, the pH of rain can change significantly during the course of a rain event, so it can be interesting to collect samples at intervals. For rain lasting one or several days, the container might be changed every few hours or twice a day. Alternatively, the container might be changed every 15 minutes during an afternoon thunderstorm. However you collect your samples, it is important to label each container clearly with the date, time of day, total sampling time, and location where it was collected.

Samples should be brought into the laboratory for measurement as soon as possible, and filtered into another clean bottle to remove any dust and debris that might react with the acids in the rain. If filtration and pH measurement cannot be done immediately after collection, samples should be stored in the refrigerator until they can be brought to the lab.

Part III. Measuring pH of Rain Samples

The pH of the rain samples will be measured using a pH meter. Refer to the *Laboratory Methods* section of this lab manual for instructions of how to operate a pH meter. It must first be calibrated using standard buffer solutions of pH 4 and pH 7. When calibrating, gently stir or swirl the buffer solution with the electrode for 30 seconds, and then let the electrode stand undisturbed in the buffer for 30 seconds. Adjust the meter as needed to the correct pH.

Rinse the electrode thoroughly with distilled or deionized water between each sample, and gently blot it dry with a tissue. Insert the electrode into a clean beaker containing 10-20 mL of the rainwater sample. Stir or swirl the solution to ensure homogenous contact with the electrodes, and then allow the solution to settle for approximately 30 seconds. Record the pH after the reading has stabilized. Record all of your pH values in your data table.

Clean up

Dispose of your samples in appropriate waste containers as instructed. Clean your pH electrode and leave it soaking in water. Clean all glassware by first rinsing with tap water and then with distilled or deionized water.

Analyzing Evidence

1. Describe the collection site(s) for your samples, and briefly explain the predictions you made about the relative acidity of the samples.

2. Did the pH of your samples match what you predicted? Explain.

3. Collect the class data and classify each of the samples according to the acid rain scale in the table at the beginning of the investigation.

Interpreting Evidence

1. If several samples from one rain event are available, did the pH change during the event? Try to provide an explanation for any changes you see.

2. If the samples were collected at different locations, are there any pH differences between locations? If so, how can you explain them?

Making Claims

What can you claim about how location and timing affect the pH of rain?

Reflecting on the Investigation

1. Based on your measurements, do you live in an area affected by acid rain? If so, what effects do acid rain have on your community?

2. Sometimes, rain samples are found to be not acidic at all, and instead are basic with a pH greater than 7. What might account for this? Formulate a hypothesis. What chemical tests or investigations could be performed to support or refute your hypothesis?

Investigating Solubility

Asking Questions

- How do scientists develop a procedure for measuring properties of materials in the laboratory?

- What data is important when defending a method for solving a problem?

Preparing to Investigate

This laboratory exercise is a departure from the usual investigation. It will simulate the kind of problem solving that takes place in a scientific laboratory. There are no instructions or procedures. There is simply a problem to solve, which requires reasoning skills and the application of knowledge that has been previously acquired. You will work in a team of three or four members.

A variety of materials and equipment will be available that you can use to solve the problem. At a minimum, you should have access to graduated cylinders, burets, beakers, flasks, plasticware, stirrers, test tubes, plastic pipets, and an analytical balance. You may wish to review the *Laboratory Methods* section at the beginning of the laboratory manual to review how to use some of this equipment.

Making Predictions

First, assign roles to each person in your group, including someone to record your ideas, procedures and results. After reading *Gathering Evidence*, develop a method to solve the problem, and then perform your investigation and record your results. Be prepared to compare your method and results with those of the other teams in the class.

Gathering Evidence

Your task is to determine the **solubility** of each of the following compounds in water at room temperature:

- calcium sulfate – $CaSO_4$

- potassium aluminum sulfate – $KAl(SO_4)_2$

- potassium nitrate – KNO_3

- ammonium nitrate – NH_4NO_3

- copper(II) nitrate – $Cu(NO_3)_2$

- sodium chloride – $NaCl$

Solubility is defined as the maximum mass of solid that can dissolve in a given volume of water. It is usually reported as grams of solid per mL of solution, or grams of solid per 100 mL of solution. Think carefully about how the procedure you design will help you measure this information.

All of these compounds, except the copper salt, are likely to be found in drinking water supplies, because they occur naturally in rocks and soil or are used as components in fertilizers. Copper(II) nitrate is included because copper salts are often used to reduce algae in ponds and swimming pools, and so they can also be found in the water supply.

In the interest of cost and convenience, you may use no more than 2 g of each compound. Your group can use graduated cylinders, burets, beakers, flasks, plasticware, stirrers, test tubes, plastic pipets, and an analytical balance.

Your team should be prepared to defend your answer and the method you used to obtain it. Therefore, it is important for your team to keep a complete record of everything you do and the numerical data you obtain. It also is important that you present the data to provide evidence that your method solves the presented problem. Focus on the quality of the procedure you develop and the care with which you carry out your investigation.

Analyzing Evidence

After you perform your investigation, rank the compounds from least soluble to most soluble.

Interpreting Evidence

1. Compare your methods and results to those of your classmates. How well did your procedure work? Can you think of any modifications or improvements you would implement if you were to do this type of investigation again?

2. Compare your results to the known solubility of each compounds. Did you rank the compounds in terms of their solubility correctly?

Making Claims

What can you claim about the process of science?

Reflecting on the Investigation

What insights do you have into how scientists design investigations and work collaboratively?

Polymer Synthesis and Properties

Asking Questions

- What is a polymer?

- How does the molecular structure of a polymer affect its properties?

- What are some common items that are made of polymers?

- What properties of a polymer would make it suitable for use in a water bottle? As fibers for clothing? As a component of a motorcycle helmet?

- What polymers are found naturally in living things?

Preparing to Investigate

Polymers are long molecular chains composed of smaller repeating units called **monomers**. Natural polymers such as proteins, DNA, starch, and cellulose are important to all living systems. Many common household items you use are made of synthetic polymers such as polystyrene and nylon. Although the polymers that we will study are composed primarily of carbon, their properties vary. We can make polymers that are opaque or transparent, rigid or flexible, weak or strong, sticky or smooth. In this investigation, you will examine the properties of several polymers, including some that are commercially available and others that you will synthesize. You will also relate the properties of the polymers to their molecular structures.

The properties of a polymer depend on the nature of its monomer and the overall chemical structure of the polymer, including chain length. In addition, the arrangement of the polymer chains relative to each other plays a large role in the material properties. In most polymers, the molecules are arranged randomly (Figure 23.1a). The mechanical strength of such polymers will be equal in all directions. In other polymers, however, the molecular chains are arranged in parallel to each other (Figure 23.1b). The strength of a polymer with parallel chains varies with direction. Pulling lengthwise meets with resistance from the covalent bonds within a chain. Pulling crosswise, however, has less resistance and simply moves one chain further away from another.

a.   Observe and record the properties of this polymer. Test the strength of the polymer by gently pulling first lengthwise and then side-to-side. Record your observations.

b.   Now grasp the film by its shorter sides. Pull firmly and slowly until the middle of the strip has stretched significantly. Observe and record the properties of this stretched section. Compare the width of the stretched section to that of the unstretched end portions. Test the strength of the stretched portion both lengthwise and crosswise. Record your observations.

3. **Mater-Bi™**. Obtain a piece of Mater-Bi™ film that has been cut from a BioBag. Mater-Bi™ is a natural and renewable polymer formed by crosslinking amylose, a linear form of starch, to varying degrees to obtain polymers with different properties. The piece you have been given is from a bag used to collect food waste for community composing projects. Test the properties of the Mater-Bi™ as you have the previous two polymers and record your observations.

B. Degradation of polymers

Place a small piece of each of the three polymers into separate wells of a wellplate. Cover each sample with enough 3-M H_2SO_4 (sulfuric acid) to cover the polymer. Let the polymers sit in the acid for at least 45 minutes while you conduct the rest of the investigation. After this time, carefully remove the polymer pieces from the wells and rinse with water. Re-examine the properties of the polymers.

CAUTION! Sulfuric acid is corrosive. Always wear eye protection and keep the solution off your skin. If you spill any acid on your skin, make sure to rinse with plenty of water and let your instructor know what happened.

Part II. Synthesizing Nylon

Nylon is a very strong polyamide fiber described in Chapter 9 of *Chemistry in Context.* To prepare nylon, you will carry out a condensation reaction of a diacyl chloride with a diamine, as shown below.

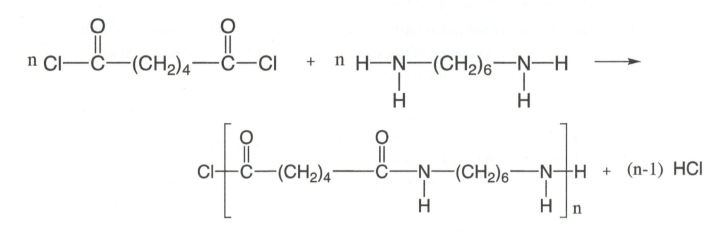

In this reaction equation, n and n-1 represent the molar equivalents of each compound. In the structure of the nylon polymer, the n subscript indicates the number of repeat units (shown within the brackets) in the polymer chain.

1. Place 10 mL of nylon solution A (adipoyl chloride in hexane) in a 50-mL beaker. Place 10 mL of nylon solution B (hexamethylenediamine in water) in a second 50-mL beaker.

     CAUTION! Wear gloves when doing this investigation, and be careful not to touch the nylon with your bare hands until it has been thoroughly rinsed with water. The monomers are hazardous and the hydrochloric acid formed as a reaction by-product can cause chemical burns.

2. Tip the beaker containing solution B at a slight angle and *slowly* add solution A by pouring it gently down the side of the beaker. Solution A should form a separate layer on top of solution B. *Do not stir or mix.*

3. A film of nylon will form between the layers. Use tweezers to grasp the film and gently pull it up and out of the beaker. Wrap the film around a test tube and rotate the test tube to gradually pull the rest of the filament from the beaker. If you are careful, you can get one long continuous strand of nylon. If the strand breaks, grab it again with tweezers and wrap it around the test tube again. Continue pulling the nylon film until there is no more solution in the beaker.

4. Rinse your nylon thoroughly under a gently stream of tap water. Slide the nylon off of the test tube. Observe and record the properties of the wet nylon.

5. Spread some of your nylon strands on a paper towel to dry. Continue with other parts of the investigation but return to observe and record the properties of the nylon after it dries.

6. Dispose of the solid nylon and any excess solutions as directed by your instructor.

Part III. Synthesis of Polyurethane Foam

A urethane is a functional group formed from the reaction between the functional groups isocyanate and alcohol. The formation of linear polyurethane from a diisocyanate with a diol depicted below is an example of urethane formation.

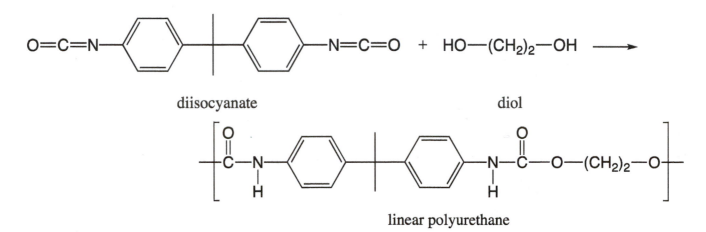

diisocyanate   diol

linear polyurethane

To form the **cross-linked** polyurethane foam, one or both of the monomers contain more than two functional groups. Additionally, a small amount of water is added to react with the isocyanate and form carbon dioxide gas. This reaction causes the polyurethane to foam, and surfactants and other additives enhance foam formation. The chemistry is complex, but the result is a strong and rigid, but lightweight, material that is used for applications requiring thermal insulation or impact resistance. Your instructor may give you additional information about the reaction that you will be performing.

1. Place 4 mL of polyurethane solution A (which contains the alcohol monomer) in a 3-oz. paper cup. Be patient as you pour so that you get as much of the viscous liquid into the cup as possible. Add 2 drops of food coloring. Mix with a wooden splint or disposable applicator stick.

2. Add 4 mL of polyurethane solution B (which contains the isocyanate monomer) to the paper cup. Use the wooden splint to *thoroughly* mix the liquids. Once the liquids are well mixed, remove the mixing stick and allow the cup to sit undisturbed.

3. The reaction may take a minute or two to begin. Once it does, record your observations. You may gently touch the outside of the cup as the reaction proceeds to observe any change in temperature, but do not touch the polymer itself yet.

4. The reaction will be complete after about 5 minutes. The polymer should no longer be sticky and can be touched. Observe and record the properties of the polyurethane foam.

5. Dispose of the polyurethane foam as directed by your instructor.

Part IV. Synthesis of Polyvinyl Alcohol Gel and Viscosity Measurements

Polyvinyl alcohol (PVA) contains a long chain of carbon atoms with alcohol functional groups attached on alternating carbon atoms. When sodium borate ($NaB(OH)_4$) is added to PVA, it reacts with the alcohol groups to make covalent bonds between polymer chains and results in the formation of a cross-linked polymer, as shown in the reaction below:

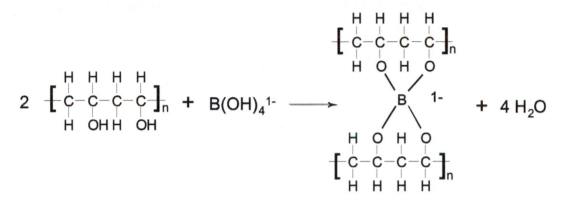

Polymers with more abundant or stronger cross-links exhibit more rigidity or **viscosity**. You will compare the properties of the original polymer to samples with varying degrees of cross-linking.

A. Synthesis of PVA gel

1. Place 50 mL of 4% PVA solution in a 100-mL beaker. Note that pure PVA is a white solid, but it has been dissolved in water for this reaction. Observe and record the properties of the polymer solution.

2. Add 5 mL of 4% sodium borate solution to the PVA solution while stirring. Continue stirring for several minutes until the mixture becomes homogenous. You may wish to use a metal spatula rather than a glass rod to stir because the mixture becomes very thick.

3. When the reaction is complete, remove the polymer from the beaker and examine it. You can touch the polymer with your hands, but wash your hands afterwards. Try forming a

ball with the polymer. Does it bounce? Does it remain round? Does it stretch? Observe and record the properties of your PVA gel.

4. Dispose of the PVA gel as instructed.

B. Viscosity measurements

In this portion of the investigation, you will determine how the amount of sodium borate solution added to PVA affects the viscosity of the resulting cross-linked polymer. Your instructor will assign you a specific amount of sodium borate to add, and you will combine your results with the class to fully analyze your findings.

1. Place 50 mL of 4% PVA solution in a 100-mL beaker.

2. Place your assigned volume of 4% sodium borate solution (0 to 5 mL) in a small beaker. Add water to bring the total volume of your solution to 5 mL.

3. Mix the two solutions, and record the time when you added the sodium borate to the PVA. Stir with a metal spatula and break up any solid lumps. Periodically mix and break up lumps for 5 minutes so that you have a homogenous gel.

4. Transfer your gel to a 50-mL beaker. Use a ruler and a marker to draw two lines exactly 30 mm apart on the side of the beaker. The top line should be somewhat below the surface of the gel and the bottom line should be somewhat above the bottom of the beaker (shown).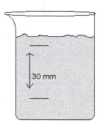

5. Hold a steel ball above your gel so that it just touches the surface. Release the ball and record the time in seconds that it takes the ball to move from the top line to the bottom line.

6. Share your data, including the volume of sodium borate used and the time for the ball to drop, with the rest of the class, and record the other data for analysis.

7. Dispose of the gel mixture as directed by your instructor.

Analyzing Evidence

Part I

1. Which polymer was the strongest? Which was the weakest?

2. How did the strength of the Teflon™ and stretched HDPE change with direction?

3. Were any of the polymers affected by reaction with sulfuric acid? How did the properties change? Which of the polymers seemed to be most resistant to the reaction with sulfuric acid?

Part II

4. How did the lengthwise strength of your nylon strand differ from its crosswise strength?

5. How did the properties of the wet nylon differ from those of the dry nylon?

Part III

6. How do the properties of your polyurethane foam differ from Teflon™, HDPE, Mater-Bi™, and nylon?

Part IV

7. Compare the properties of the PVA solution to the PVA after adding the sodium borate solution. How is the substance different?

8. How is the viscosity of the PVA affected by adding different amounts of sodium borate? Make a graph of the class data, recording the time for the steel ball to drop (in seconds) on the *y*-axis versus the volume of sodium borate used on the *x*-axis. Can your data points be connected by a straight line, or does the relationship seem to be represented by a curved line?

Interpreting Evidence

1. Do you think the polymer strands in Teflon™ tape are arranged randomly or in parallel? Explain your reasoning.

2. When HDPE is stretched, a "neck" can form. What do you think happens to the arrangement of the polymer strands in this region of the polymer? Support your answer with experimental data.

3. Nylon is often drawn into fibers. What lengthwise and crosswise strength do you think would be important for a polymer fiber to exhibit? Explain your reasoning. Did your observation of nylon's properties match this expectation?

4. When you made nylon, you were warned that the monomers can be hazardous but that the polymer, once rinsed, is relatively harmless. Speculate on the difference in hazard between the monomers and the polymer.

5. Both polyurethane foam and polyvinyl alcohol gel contain cross-links. What are cross-links? How do the strengths of cross-links in these two polymers compare to each other? Back up your assertion with experimental evidence.

6. Describe the relationship between the viscosity of PVA gel and the amount of added sodium borate. Explain the connection to crosslinking.

Making Claims

What can you claim about how polymer properties are affected by the molecular structure of the polymer? Compare polymers based on chemical composition, cross-linking, and other structural features.

Reflecting on the Investigation

1. Mater-Bi™ is one example of the new biodegradable polymers that have come on the market in the last decade. Look online for answers to these questions. Cite your sources.

    a. Identify two other commercially available biodegradable polymers.

    b. Besides their use in the collection of waste for community composting, what other applications do biodegradable polymers have? (One example: Have you ever had dissolving stitches to close a wound?)

    c. Polyurethane solutions A and B are sometimes injected directly into the walls of older homes and allowed to foam in place. Why?

2. Teflon™ is commonly used as a nonstick coating on cookware. A compound used in its processing, perfluorooctanoic acid (PFOA), has been a source of public controversy. Look online for answers to these questions. Be sure to cite your sources.

    a. List two concerns about PFOA.

    b. How might the key ideas in green chemistry (see p. viii) be used to address these concerns?

Identifying Common Plastics

Asking Questions

- What do you use in your real life that is made of plastic?
- What common types of plastic are present in the items that you use often?
- Which of these items are easy to recycle in your area, and which are more difficult?
- What are the six commonly recyclable plastics designated by recycling symbols?
- Why must plastics get sorted for recycling?
- How can the properties of the plastics be used to sort them?

Preparing to Investigate

Plastics have been an important part of our industrial society since Leo Baekeland (1863–1944) invented the first synthetic polymer in 1907. It was a thermosetting resin called Bakelite, which offered an alternative to cellulose resins. Today, plastics are found in everything from food and beverage containers to furniture and electronic equipment. According to the most recent data available from the U.S. Environmental Protection Agency (EPA), about 254 million tons of municipal solid waste was generated in the U.S. in 2013. This waste contained 32 million tons of plastics, including 14 million tons of containers and packaging. Of all these plastics, a mere 3 million tons, about 9%, is recycled. This rate is well below those for other materials such as newspapers (67%), aluminum cans (55%), and glass containers (34%).

The majority of plastic waste is composed of six common polymers, "The Big Six." The plastics industry has adopted a code for packaging materials that can be used to identify the type of plastic in a container. The idea behind the symbol is to make recycling easier by making the identification of the plastics more straightforward. The symbol on the bottom of many containers is a triangle of arrows with a number in the middle of the triangle. Table 24.1 lists the type of plastic represented by each number.

Compliance in labeling is voluntary, and not all plastics have an identification code symbol. Without code numbers, plastics are difficult to separate by appearance.

Table 24.1. Plastic recycling numbers.

Number	Name	Abbreviation
1	Polyethylene terephthalate	PET
2	High-density polyethylene	HDPE
3	Polyvinyl chloride	PVC
4	Low-density polyethylene	LDPE
5	Polypropylene	PP
6	Polystyrene	PS
7	Other (includes composites and mixtures)	

In this investigation, you will examine the properties of six types of plastics and develop a classification and identification scheme. You will perform the following tests:

1. The relative **density** of each plastic will be measured by checking to see whether the samples float or sink in three liquids of differing densities.

2. All six of these common plastics melt reversibly, which means that when they are cooled, they harden and may regain their original properties. If a plastic sample does not melt, it is a **thermosetting plastic**. Thermosetting plastics, such as melamine (used in high-quality plastic dinnerware), do not melt cleanly and reversibly but tend to char instead.

3. All common plastics burn (some only if held directly in the flame), but they do so with slightly different characteristics and different noxious fumes. The vapors from the burning plastic may have different properties depending on the plastic. The ignition test must be performed only in a **fume hood**.

4. To determine whether a halogen, such as chlorine, is part of the polymer, you will use a test called the copper-wire test.

Making Predictions

After reading *Gathering Evidence*, prepare a table to record your investigation results from the four tests on known plastics. After you have devised your classification scheme, you will need to prepare another table for your results on the unknown substances. Which polymer should give a positive copper-wire test?

Gathering Evidence

Overview of the Investigation

1. Obtain known samples of the six plastics and perform four tests on each sample.
2. Devise a classification scheme and a way to identify each plastic.
3. Test your classification scheme with known samples.
4. Identify unknown samples using the four tests and your classification scheme.

Part I. Density Tests

Three liquids of differing densities will be used, as shown in the following table:

Liquid	Density (g/mL)
47.5% ethanol (*aq*)	0.94
Water	1.00
10% NaCl (*aq*)	1.08

1. Obtain three test tubes and label them with the identity and density of each solution. Put about 5 mL of each liquid in the test tube labeled with its density.
2. Place two narrow strips of each of the six types of plastic in a labeled vial or beaker. Cut *one* strip of each plastic into three small pieces and store in the labeled container.
3. For the first plastic sample, place one piece into each of the three test tubes containing the density test liquids. Push each piece under the liquid surface with a glass stirring rod. If the sample floats, it has a density lower than that of the liquid. If it sinks, it has a density higher than that of the liquid. Record your observations in your data table.
4. Remove the plastic samples from the test tubes with a pair of tweezers, and then test each of the other plastics one at a time. Record your observations.

Part II. Melt Test

1. Place a small sample of each plastic, one at a time, on the end of a metal spatula and hold the end of the spatula over a light blue burner flame.

     CAUTION! Take great care when using an open flame. Long hair must be tied back, and loose sleeves should be secured. Pieces of hot molten plastic can cause burns if dropped onto your skin or clothing. Additional information on using a burner can be found in the *Laboratory Methods* section.

2. Heat slowly and observe the plastic as it warms and finally melts. DO NOT heat the sample so strongly that the plastic catches on fire. Record your observations in your data table.

3. Cool the sample and examine its appearance and observe its flexibility by bending it. Record your observations. You may use your melted and cooled plastic for the subsequent tests.

Part III. Ignition Test

 CAUTION! Toxic fumes are produced during the ignition test, so it is imperative that this part of the investigation be done in a fume hood. If a fume hood is not available, do not do this part of the investigation.

1. Place a Bunsen burner and a large beaker of water in a fume hood. Light the burner and adjust it to a small flame.
2. Hold one end of a small strip of plastic with a pair of tongs, forceps, or pliers and place it directly in the flame. Observe the color of the flame and its characteristics. Is a lot of smoke or visible vapor given off? Does the plastic continue to burn after it is removed from the flame? Record your observations.
3. Test the vapors given off for acidic properties by holding a piece of *wet* litmus paper in the vapors above the burning plastic. If the paper turns red, acidic fumes are being formed as the plastic burns. Record your observations.
4. Extinguish the burning plastic by dropping it into the beaker of water. Repeat the ignition test for the other plastics.

Part IV. Copper-Wire Test

CAUTION! Toxic fumes are produced during the copper-wire test, so this part of the investigation MUST be done in a fume hood. If a fume hood is not available, do not do this part of the investigation.

1. Push the end of a 6-inch length of copper wire into a small cork.
2. Use the cork as a handle and heat the free end of the wire in a burner flame until the flame has no green color.
3. Touch the hot copper wire to the plastic you are testing and then return the wire end to the flame. The hot wire should pick up a tiny bit of plastic. Return the wire end to the flame. When the tip of the wire is put in the flame, watch for a slight flash of luminous flame. This indicates that you have correctly picked up a little bit of plastic on the wire.
4. Watch for the appearance of a green flame or green color in the flame when the plastic is heated in the flame. The green color indicates the presence of a halogen, such as chlorine, in the plastic.
5. Test each plastic sample and record your observations.

Part V. Identifying Unknown Plastics
1. Work ahead to the *Analyzing Evidence* section. In that section, you will analyze your results from the four tests and use this data to devise a method for identifying the six plastics that you were given.
2. After completing the work in *Analyzing Evidence*, use your classification scheme to identify at least two plastic samples that you brought from home or that your instructor has provided for you.
3. As a third step in the investigation, two unknown plastic samples will be given to you to identify. Use your scheme to determine their identities.

Analyzing Evidence

1. Rank your six plastics from least dense to most dense.
2. What differences did you observe between the plastics in the melt test?
3. What differences did you observe in the ignition test? Did one plastic stand out? If so, describe this result.
4. What differences did you observe in the copper-wire test? Did you correctly predict which plastic(s) would give a positive halogen test?
5. Devise a scheme for identifying the six plastics based on the simple tests you have performed. The challenge is to find the minimum number of tests that will correctly identify the plastics when you are given any one of them as an unknown. Draw a flow chart or outline your identification scheme, then return to Part V of *Gathering Evidence*.

Interpreting Evidence

1. Were all four of the tests necessary for identifying the six plastics, or could you have focused on fewer tests?
2. Suppose you had to add two other plastics to your scheme, polymethylmethacrylate (density 1.18–1.20 g/mL) and poly-4-methyl-1-pentene (density 0.83 g/mL). Where would they fit into your scheme?

Making Claims

What can you claim about how properties of plastics can be used to identify them? Did you devise a reasonable method for identifying the different plastics?

Reflecting on the Investigation

1. Why are plastic recyclers concerned about identifying the different polymers and not mixing them together?

2. Polyethylene terephthalate (PET) is the most valuable waste plastic at the present time. Suggest a way to separate it on a commercial scale from other waste plastics.

3. Since waste plastic consists mostly of hydrocarbon compounds, it has been suggested that waste plastic could be used as fuel. Based on your observations in this investigation, do you think this is a reasonable suggestion? Would some plastics be more dangerous to burn than others? Defend your answer.

Degradation of Poly(lactic acid)

Asking Questions

- What chemical reactions can we use to degrade polymers?
- How can we use chemistry to take abundant feedstocks and convert them into useful products?
- What aspects of poly(lactic acid) would support the idea that it is an environmentally friendly polymer?
- How does surface area impact the rate of reactions?

Preparing to Investigate

In this investigation, a **polymer** will be degraded into its **monomer** units through a process called depolymerization (see Investigation 23 and Chapter 9 in *Chemistry in Context*). The polymer in post-consumer waste poly(lactic acid) (PLA) cups will be converted to lactic acid, an anti-microbial cleaner, through base hydrolysis.

PLA is a biobased and biodegradable polymer. Unlike traditional plastics that are derived from petroleum sources, PLA is produced from plant carbohydrates including those from corn or sugarcane. Most polymers degrade very slowly, but the structure of PLA enables composting enzymes and bacteria to break it down in about three months. The rate of biodegradation depends on environmental conditions, so PLA may break down more slowly in a backyard composter than it does at municipal and industrial composting facilities.

PLA is used widely in food and beverage packaging, medical devices, and 3-D printing. PLA is an easily processed material; therefore, many products can be made from this polymer, including spinnable fibers for making textiles.

To convert PLA into lactic acid (Figure 25.1), the polymer is hydrolyzed by sodium hydroxide, NaOH, and then acidified with hydrochloric acid.

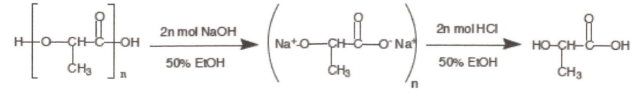

Figure 25.1. Basic hydrolysis of poly(lactic) acid.

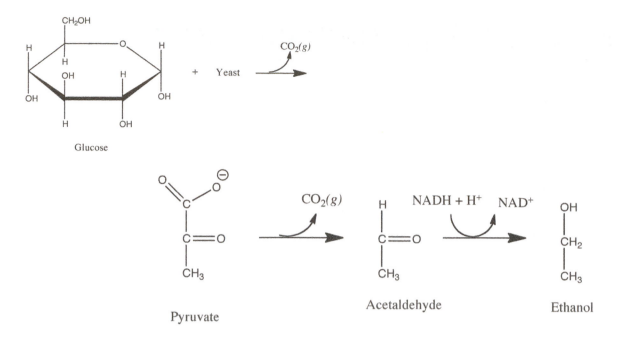

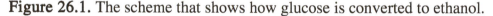

Figure 26.1. The scheme that shows how glucose is converted to ethanol.

glucose to ethanol and carbon dioxide (Figure 26.1). If oxygen is present, the ethanol can be oxidized to acetic acid. In an oxygen-free environment, one mole of sucrose produces four moles of ethanol and four moles of carbon dioxide. This reaction does not go to completion, though, because the yeast die when the alcohol concentration reaches somewhere between 10% and 20%.

The boiling point of a liquid is the temperature where the vapor pressure of the liquid is equal to the atmospheric pressure. The liquid rapidly vaporizes at this temperature. The process of distillation takes advantage of this behavior to separate mixtures of liquids. If liquids have different boiling points, they will turn to vapor at different temperatures. By vaporizing and condensing a liquid at a particular temperature, you can isolate a single substance. Simple distillation is effective for separating liquids that have significantly different boiling points. In fractional distillation, a fractionating column that has many surfaces for condensing vapor is added. This process can separate liquids with much more similar boiling points.

The boiling points of ethanol and water are separated by more than 20 °C, but pure ethanol is still very difficult to obtain. Ethanol and water form an azeotrope (or constant boiling mixture) of 96.4% ethanol and 4.4% water, which boils at 78.2 °C. Since the boiling point of the azeotrope is lower than either of the liquids in the mixture and very close to that of ethanol, water is almost always present in distilled ethanol. Because water and ethanol have different densities, the ethanol content in the distillation product can be determined by measuring its density.

Please note that during this investigation you will be distilling a product that could be called rum or bioethanol fuel. It will not have a desirable flavor and is not fit for human consumption. In addition, the conditions in the lab could lead to unintended compounds being dissolved in the ethanol because it is a good solvent. Our interest in this investigation is in the synthesis and isolation of an organic compound, ethanol.

Making Predictions

- After reading *Gathering Evidence*, prepare a data sheet to record your observations during fermentation, after fermentation, during distillation, and after distillation. Include a data table where you can record the densities before and after distillation, times during the distillation, and boiling points.
- Find the densities and boiling points for ethanol and water and list them at the top of your data table.
- What would happen if air entered the flask during fermentation? How are you preventing this from happening during this investigation?
- Predict the products that will be in the flask after fermentation and before distillation. Predict the products that will be in the flask after distillation.
- Why is it impossible to distill pure ethanol from a reaction mixture that includes water?

Gathering Evidence

Overview of the Investigation
1. Prepare a mixture of molasses, water, and yeast.
2. Set up a fermentation flask with a one-way vapor lock and store for one week.
3. Decant ethanol solution and measure density.
4. Perform a simple distillation.
5. Determine the ethanol content of your final product.

 CAUTION! Never eat or drink in the lab, and do not eat or drink anything that has been created in the lab.

Part I. Making Ethanol through Fermentation
1. Place 70 mL of water in a 250-mL Erlenmeyer flask.
2. Add 70 mL of blackstrap molasses and stir with a glass rod until completely mixed.
3. Add about 0.5 g of yeast and stir again. The yeast will not dissolve and may float on the surface of the molasses solution.

4. Put a bent glass tube through the hole of a one-hole rubber stopper.

    CAUTION! Do not force the glass tube into stoppers or rubber tubing. Lubricate the glass with soapy water. Use cloth to hold the glass and the stopper or tubing. Gently twist the glass tube into the hole.

5. Attach a short piece of rubber tubing to the bent glass tube. Put a straight glass tube or glass pipet into the other end of the rubber tubing.

6. Add 100 mL of limewater (aqueous solution of calcium hydroxide) to another 250-mL Ehrlenmeyer flask.

7. Stopper the flask and make sure that the stopper is secure.

8. Put the straight glass tube or pipet in the other flask. Make sure the tip of the glass is under the surface of the limewater. This makes a one-way vapor lock so that oxygen cannot get into the flask, but carbon dioxide can get out.

9. Record your observations about the fermentation and the date and time that fermentation began. Store your flasks securely for one week.

Part II. Distilling Ethanol

1. Your instructor will show you how to set up a simple distillation using the equipment that is available in your laboratory. Follow their instructions and do not proceed until your apparatus has been checked for correctness and safety.

2. Record your observations about the fermentation reaction and the date and time that fermentation was ended.

3. Decant the ethanol solution into a 250-mL round-bottom flask.

4. Measure the mass of a small vial or beaker. Add 5 mL of your solution and measure the mass again. Subtract to determine the mass of the solution.

5. Distill the solution. Collect the alcohol fraction until the temperature is just below the boiling point of water. At that temperature, remove the heat source and replace the collection flask with an empty one. Record times, temperatures, and observations throughout the distillation.

6. Measure the mass and the volume of your distilled product.

Clean up

Discard or recycle the contents of your test tubes as directed by your instructor. Clean your thermometer with soap and water to remove all traces of the oil.

Analyzing Evidence

1. What was the range of boiling points during collection of your distillation product?
2. Calculate the density of your ethanol solution before and after distillation.

Interpreting Evidence

1. You used a one-way vapor lock during fermentation that included limewater. Limewater reacts with carbon dioxide to form calcium carbonate, a white solid that is insoluble in water. Calcium carbonate can react with carbon dioxide to form calcium bicarbonate, a water-soluble compound. How does this information support an argument that air was excluded from your fermentation reaction?

2. You made predictions as to what products would be in your flask after each process. Were your predictions correct? How do you know?

3. How did the density of the solution change from before distillation to after distillation? Does this suggest that there is a larger alcohol content before or after distillation? Why?

4. Use Table 26.1 to determine the ethanol content of your ethanol solution before and after distillation. Does this data support or refute your answer to Question 3?

Table 26.1. Aqueous Ethanol (EtOH) Content

Density g/mL	% EtOH by wt.	% EtOH by vol.		Density g/mL	% EtOH by wt.	% EtOH by vol.
0.989	5	6.27		0.856	75	81.30
0.982	10	12.44		0.843	80	85.49
0.975	15	18.54		0.831	85	89.48
0.969	20	24.54		0.828	86	90.25
0.962	25	30.46		0.826	87	91.02
0.954	30	36.25		0.823	88	91.77
0.945	35	41.90		0.821	89	92.53
0.935	40	47.40		0.818	90	93.27
0.925	45	52.72		0.815	91	93.99
0.914	50	57.89		0.813	92	94.72
0.903	55	62.89		0.810	93	95.44
0.891	60	67.74		0.807	94	96.11
0.880	65	72.43		0.804	95	96.79
0.868	70	76.95		0.789	100	100.00

Making Claims

What can you claim about the fermentation of molasses and distillation of ethanol?

Reflecting on the Investigation

1. What would have been the advantage of doing a fractional distillation after the simple distillation?
2. Making ethanol through the fermentation of sugars is often identified as being a green process. Use the key ideas of green chemistry (p. viii) to either defend that argument or to support the counterargument.
3. Do you know for sure that your final product contained or did not contain some of the ethanol/water azeotrope? Give evidence to support your answer.

Investigation adapted from Thompson, J. Biosynthesis of Ethanol from Molasses.

Protein in Milk and Other Foods

Asking Questions

- What is a protein? What molecules combine to form proteins? What are the linkages between those smaller molecules called?
- Describe the difference between the primary and secondary/tertiary structure of proteins.
- Of the foods you ate yesterday, which contain the most protein? Which have the least?

Preparing to Investigate

Proteins are composed of **amino acids** linked together by amide bonds into long polymer chains. Structural proteins such as collagen and keratin are important components of the skin, hair, feathers, tendons, and bones of animals, while enzymatic proteins catalyze biochemical reactions throughout the plant and animal kingdoms. Proteins can be described by their **primary structure**, which is the order of the amino acids along the chain. Additionally, proteins usually assume an overall complex three-dimensional shape, the **tertiary structure**, through interactions of the amino acid side-chain functional groups. Changes to either the primary or tertiary structure of a protein can drastically affect its properties, including strength, reactivity, and solubility.

Twenty naturally occurring amino acids are required to build the proteins in our bodies. We can make 11 of these from other components of our food, but nine of them are considered "essential" – we must eat them in order for our bodies to have them available for biosynthesis. Meat, dairy products, and eggs contain all of the essential amino acids, so we consider them to be "complete" proteins. Vegans, who don't eat any animal products, have to make sure they consume all of the essential amino acids, which are found in many plant-based foods as well.

In this investigation, you will isolate proteins in the form of curds from cow's milk and soy milk by **precipitation**, and will also test foods for protein content. The main protein in cow's milk, **casein**, can be precipitated in two ways. First, you will use vinegar, which is acidic, to lower the pH of the milk, causing the casein to change how it is folded and precipitate out of solution. This is the process by which fresh cheese, such as ricotta, is made. The second method uses rennet, an extract containing the enzyme rennin that cleaves some of the peptide linkages in the casein and causes it to become insoluble. Hard cheeses, such as cheddar, are made using this method. For soy milk, you will use Epsom salt ($MgSO_4$) to precipitate the main proteins in soy, glycinin and β-conglycinin. Tofu is made this way. You will test your curds and other foods using the Biuret test, which causes a color change in protein-rich foods.

Making Predictions

After reading *Gathering Evidence*, create some data tables to record your experimental results. Which type of milk do you expect to have more protein–soy milk or cow's milk? Should vinegar or rennet yield a higher amount of protein precipitate when reacted with cow's milk? Which of the foods in Part II do you expect to contain the most protein?

Gathering Evidence

Overview of the Investigation

1. Precipitate protein from cow's milk and soy milk
2. Test for the presence of protein in your cheese curds and other foods using the Biuret test.

 CAUTION! Do not consume any of the food products used in the laboratory!

Part I. Precipitation of Proteins

A. Acidic precipitation of milk casein

1. Record the mass of an empty 250-mL beaker. Then, measure 120 mL of pasteurized whole milk into the beaker using a graduated cylinder. Weigh the milk plus the beaker, and calculate the mass of the milk.
2. Heat the milk on a hot plate to 20°C, then remove the beaker from the hot plate.
3. Measure and add 10 mL of white vinegar to the milk. Stir it with a glass rod for 2 minutes, and then let the beaker sit undisturbed for 5 minutes while the proteins precipitate.
4. Cut 2-3 layers of cheesecloth to fit over the mouth of the beaker, and fasten them onto the beaker with a rubber band. Pour the mixture through the cheesecloth, so that the curds will be collected in the cloth and the remaining liquid in the beaker.
5. Rinse the curds, squeeze them dry, and spread them out on a piece of absorbent paper (such as filter paper or a thick paper towel). Allow them to dry for at least 5 minutes.
6. Measure the mass of the curds, record your observations about their appearance, and save them in a beaker for the Biuret test.

B. Enzymatic precipitation of milk casein

1. Crush a tablet of rennet by putting it between two pieces of aluminum foil and smashing it with a hammer. Record the mass of an empty 250-mL beaker. Place about half of the

tablet into the beaker. (Two groups can share one rennet tablet for this experiment.) Record the mass of the beaker with the rennet, and calculate the mass of the rennet.

2. Pour 120 mL of milk into a second tared 250-mL beaker. Weigh the beaker again and calculate the mass of the milk.

3. Heat the beaker of milk on a hot plate to 40 °C.

4. Carefully pour the warm milk over the rennet tablet, and stir the mixture with a glass rod for 2 minutes. Allow the mixture to sit for 5 more minutes as the protein precipitates.

5. Filter the curds as outlined in steps 4-6 of Procedure A.

C. Precipitation of soy milk

1. Record the mass of an empty 250-mL beaker. Then measure 120 mL of soy milk into the beaker. Weigh the milk plus the beaker, and calculate the mass of the soy milk.

2. Heat the soy milk on a hot plate to 85 °C. If the bubbling is uneven or the milk is splattering, add a few boiling chips or marbles to the beaker to promote even boiling.

3. Carefully remove the beaker from the heat using a potholder or heat-resistant gloves. Measure and add 1.6 g of Epsom salt ($MgSO_4$) to the soy milk. Stir it gently with a glass rod and then let it sit until the curds are floating in clear liquid. Stirring too vigorously or for too long will create very small curds that will be difficult to isolate.

4. Filter the curds as outlined in steps 4-6 of Section A.

Part II. Biuret Test for Proteins

The Biuret test uses a color change to indicate the presence of proteins. It is sensitive for proteins that contain three or more amino acid residues. You will test all of the curds you made in Part I, and should also select some other foods that may or may not contain significant amounts of protein. Your instructor will let you know if there will be foods available in the lab for you to test, or if you should bring some from home. *Remember to never eat any foods that have been used or stored in the laboratory!*

1. Arrange the foods that you will test in Petri dishes or on watchglasses.

2. Using a pipet, add 1 mL of 10%(w/v) NaOH solution onto the food.

CAUTION! Sodium hydroxide is corrosive and can damage skin and clothing. Wear gloves when working with it, and report any spills to your instructor immediately.

3. Using a pipet, carefully add 5 drops of 5%(w/v) solution of $CuSO_4$ onto the food, in the same location that you added the NaOH. If no protein is present, the light blue solution will stay the same color. If the tested food contains protein, the solution will become a darker violet color. Carefully record your results.

Part III. Optional Activities

1. Use vinegar or rennet to precipitate soy milk, or $MgSO_4$ to precipitate cow's milk. Follow the procedures in Part I but substitute the other type of milk. Does it work as well?
2. Try any of the three procedures with other types of milk from animals such as goats, or plant-based milks like almond, rice, or coconut milk.

Analyzing Evidence

1. Calculate the weight percent of protein curds you isolated from milk using each of the procedures.
2. Which foods have the most protein according to the Biuret test? Sort them into lists of high and low protein foods.

Interpreting Evidence

1. Milk and soy milk will list protein content on their labels. Find those values and answer the following questions.

    a. How does the percent protein isolated from cow's milk using vinegar compare to the amount of protein stated on the label?

    b. How does the percent protein isolated from soy milk compare to the amount of protein listed on the label?

    c. For both types of milk, if the isolated protein is present in a lesser amount than listed on the label, speculate how the additional protein was lost using your isolation procedure. If it is higher, what additional material may be present?

2. Define the terms *solution*, *suspension*, and *emulsion*. Which of these applies to milk? Explain your answer and back it up with evidence from your experiment or an outside source.
3. Did the Biuret test give expected results based on protein values listed on the labels of the foods you tested? Explain any anomalies.

Making Claims

What can you claim about the amount and type of proteins in milk and soy milk? Are they the same or different? What types of foods have higher amounts of protein? How do you know?

Reflecting on the Investigation

1. Why did the curds prepared using rennet have a greater mass than those prepared using vinegar, even though the same mass of cow's milk was used?

2. Describe the structure of the casein protein in milk. What would you see if you had a powerful microscope and could look at it? (*Hint:* use the Internet to learn more about casein.)

3. Vinegar works by altering the charges on the casein protein, causing it to change its shape and rendering it insoluble so that it precipitates. Does this mean that it is changing the primary or tertiary structure of the protein?

4. Rennet works by cleaving some of the peptide bonds in the casein protein. Does this mean that it is changing the primary or tertiary structure of the protein?

5. Based on your results, list some foods that vegetarians or vegans could eat to get enough protein.

6. Explain the concept of eating "complete proteins," and after doing some research list some plant-based foods or food combinations that will provide complete protein. Cite your sources.

Experiment adapted from *Food Chemistry Experiments*, Institute of Food Technologists, 2000.

Notes

Measuring Fat in Potato Chips and Hot Dogs

Asking Questions

- How much fat is in potato chips and hot dogs?
- What physical property can we use to separate the fat from foods to measure how much is present?
- What are some health consequences of excess fat in the diet?
- Do different types of chips or meat products contain different amounts of fat?

Preparing to Investigate

Fats are an essential part of our diet, but consuming too much fat has been linked to health problems including obesity and heart disease. In this investigation, you will perform an **extraction** to measure the amounts of fat in foods that are a common part of our modern diet. Extractions work by separating compounds in a mixture (such as a food) by their solubility. Two procedures will be used to determine the fat content of foods. In potato chips, the fat is mostly on the outside, so the fat can be dissolved using a suitable solvent. However, another approach must be used for hot dogs and other meat products. In meats, fat is bound to protein, so it must be chemically treated to break down the protein and liberate the fat. The fat will then float on top of water and can be skimmed off the surface easily.

Making Predictions

After reading *Gathering Evidence*, make a data table for your results. Develop several scientific questions that you can answer with this investigation, and predict the relative fat content of the samples that your class will analyze.

Gathering Evidence

Overview of the Investigation

1. Weigh food samples.
2. Grind chip samples and extract fat using petroleum ether.
3. Evaporate the petroleum ether and weigh the fat residue.
4. Mix the meat samples with protein-liquefying reagent.
5. Centrifuge meat samples, and separate and weigh the fat.
6. Calculate the percentage of fat in your samples.

CAUTION! Never consume food products that have been used in the laboratory.

Part I. Measuring Fat in Potato Chips

The procedure for extracting fat from potato chips is very simple: Petroleum ether, which is a commercial mixture of hydrocarbons that is widely used as a solvent, is mixed with ground chips to extract fat. After separating the solvent mixture from the chips, the petroleum ether is evaporated and will leave the fat that can be weighed. This procedure works well for potato chips, tortilla chips, crackers, or similar processed foods. You may wish to investigate the differences in fat content between different brands or types of chips, or assess whether products that claim to be "low fat" actually fit this description.

1. Obtain clean, dry beakers for your samples. You will need one beaker for each sample. Label the beakers so you can identify your samples.
2. Weigh the beakers and record the mass of each beaker.
3. Weigh 2–3 grams of each sample you will analyze. Record the mass to the nearest 0.01 g.
4. Use a clean porcelain mortar and pestle to crush your first sample into small pieces.
5. Add 15 mL of petroleum ether to the mortar and grind the mixture thoroughly.

CAUTION! Petroleum ether is a gasoline-like solvent that is extremely flammable. Make sure that no open flames are present anywhere in the laboratory during this investigation. Under no circumstances should the heating be done with a Bunsen burner.

6. Separate the petroleum ether from the solid chip residue by performing a **gravity filtration**. This procedure is further explained in the *Laboratory Methods* section. Prepare a glass funnel with folded filter paper and mount it over the appropriate beaker. Use a spatula to transfer the mixture into the filter.
7. Rinse the mortar twice, using 5 mL of petroleum ether for each rinse, and put the liquid into the filter to drain into the beaker.
8. Repeat the procedure for each sample. Use a clean, dry mortar and pestle and a new piece of filter paper for each sample.
9. When all samples have been extracted, you will remove the petroleum ether by evaporation. There are two ways this can be done:
   a. You can leave the beakers in a fume hood until the next day or the next lab period, and the solvent will evaporate.
   b. You can use a rotary evaporator, if one is available in your laboratory.
10. Reweigh the beakers and record the masses to the nearest 0.01 g.

Part II. Measuring Fat in a Hot Dog or Other Meat Sample

This procedure can be used for any meat sample, which allows you to compare different samples. You may wish to investigate whether different brands of hot dogs differ significantly in fat content, whether chicken or turkey hot dogs differ from those made with pork or beef, whether hot dogs and hamburgers differ significantly in fat, whether "low fat" products are really as claimed, or how vegetarian hot dogs compare to meat hot dogs in their fat content.

1. Put 75 mL of water in a 250-mL beaker. Place the beaker on a hot plate and heat the water to about 85 ° C.

2. While the water is heating, label a centrifuge tube for each sample that you plan to analyze. Weigh and record the mass of each tube to the nearest 0.01 g.

3. Put 2–2.5 g of ground meat in each test tube and weigh the tubes again.

4. Add about 5 mL of the protein-liquefying reagent to each meat sample. The protein-liquefying reagent is a solution of sodium salicylate, potassium sulfite, and sodium hydroxide in isopropyl alcohol and water.

   CAUTION! The protein-liquefying solution is caustic and can cause serious skin damage. Do not allow it to contact your skin. If you do get it on yourself, immediately wash your skin with lots of water and notify your instructor.

5. Put the centrifuge tubes in the beaker of hot water and heat until the reagent in the tube starts to boil (about 80 ° C). Maintain the temperature of the water so that the contents of the centrifuge tubes boil for 10 minutes. *Do not leave the beaker and tubes unattended.* The reaction must be monitored to ensure that liquid does not bubble out of the test tube. In addition, vapor from the protein-liquefying reagent is extremely flammable so care must be taken.

   CAUTION! Never use a Bunsen burner to heat a flammable substance. Always use a hot plate.

6. After the mixture has boiled for 10 minutes, it should be dark brown with some yellow fat floating on the top. Remove the centrifuge tubes from the hot water, stand them in an empty beaker, and let them cool until you can handle them but the fat is still liquid. Do not let them cool to room temperature, because if the fat solidifies the separation will become difficult.

7. Obtain small containers (vials, beakers, test tubes, or watch glasses) for each of your samples. Label, weigh, and record the mass of each container.

8. Put your sample tubes in a centrifuge and make sure they are balanced. Centrifuge the mixtures to separate the fat from the rest of the solution. It should be a layer on top of the solution.
9. Use a Pasteur pipet with a rubber bulb to carefully transfer the fat from the top of your sample into the appropriately labeled container. Be careful to remove all of the fat but none of the brown liquid. Work slowly; it takes patience to do this correctly so that only the fat is removed. Repeat this for each of your samples.
10. Weigh the containers of fat and record their masses.

Clean up

Discard the filter paper and chip mixtures from Part I in the designated container. The "brown protein liquid" from Part II and your isolated fats should be deposited in appropriate waste containers. Wash the mortar, pestle, beakers, and tubes thoroughly and allow them to drain.

Analyzing Evidence

To compare your samples, you should calculate the percent fat in each one. First, calculate the mass of sample (chips or meat) and the mass of fat that you isolated by subtracting the mass of the container from the mass of the container with sample or fat. Then, calculate the percent using the following equation:

$$\% \text{ fat} = \frac{\text{mass of fat}}{\text{mass of sample}} \times 100\%$$

Interpreting Evidence

1. Your instructor may ask you to share your results with your classmates. How closely do your results agree with those of your classmates for the same samples? Can you suggest any sources of error in the measurements?
2. Looking at the class results, answer the scientific questions you posed in *Making Predictions.*
3. Can you draw any generalizations about chips? For example, do potato chips have more or less fat than tortilla chips?
4. If your class analyzed different kinds of hot dogs or other meats, what can you conclude about the fat content in these products? Are some products lower in fat than others?
5. If your class analyzed any products that claim to be "low fat" or "no fat," do the results support this claim? If not, suggest a reasonable explanation.

6. A simple calculation will allow you to determine whether your results agree with the fat content reported by the nutrition label generated by the chip manufacturer. The mass of a potato chip is mostly fat, carbohydrate, protein, and water. If we assume the water content is low, dividing the reported fat content by the sum of masses of the three main components will provide you with an estimated percent fat content of the chips.

$$\% \text{ fat from package} = \frac{\text{fat content (in grams)}}{\text{sum of fat} + \text{carbohydrate} + \text{protein (in grams)}} \times 100\%$$

If the package for your potato chip sample is available, perform this calculation. Do your investigation results agree with this calculation? If not, why not? Was the assumption about water valid?

Making Claims

What can you claim about the fat content of chips and meat products? Can you draw any conclusions about the relative fat content of different kinds of foods?

Reflecting on the Investigation

1. Potatoes are naturally low in fat, so why do potato chips contain lots of fat? How do regular chips differ from "low fat" ones?

2. Why do hot dogs require a different fat extraction procedure from the one used with chips? Which of the two procedures do you think would work best when analyzing roasted peanuts, corn, or pepperoni pizza? Explain your reasoning for each food.

3. A snack pack of potato chips or other kinds of chips holds 1 ounce (28 g) of chips. Based on your data, how much fat is present in a snack pack? In such a bag of chips, there are about 15 g of carbohydrate and about 1 g of protein. Given that carbohydrates and protein provide about 4 Calories of energy per gram, and fats provide about 9 Calories per gram, what is the total energy equivalent (in Calories) of a 1-ounce bag of chips? What percent of the Calories are from fat? Show your calculations.

4. A typical hot dog weighs about 45 g. Use your data to calculate the number of grams of fat in the hot dog you analyzed. There are about 7 g of protein in a hot dog, and almost all of the remaining mass is water. Using the Calorie values given in Question 3, calculate how many Calories are provided by the fat, the protein, and the water. What percentage of the Calories in the hot dog can be attributed to the fat?

NOTES

Measuring the Sugar Content of Beverages

Asking Questions

- How does the amount of sugar dissolved in a beverage affect its physical properties?

- Can we make a device to measure the amount of sugar in beverages?

- How much sugar is in soft drinks?

- How does the amount of sugar in fruit juice compare to that in soft drinks?

Preparing to Investigate

Many of the beverages we consume each day, including fruit juices and non-milk-based soft drinks, contain sugar. Have you ever wondered just how much sugar is present in these beverages? In this investigation, you will build a simple device called a **hydrometer** that you will use to measure and compare the sugar content of a number of common beverages.

Rather than directly measuring the concentration of sugar, which would require specialized equipment, we will instead measure the **density** of the solution, a physical property that is related to the amount of sugar. The density is the mass (in grams) per volume (in cubic centimeters or milliliters) of a substance. For instance, the density of water is 1.000 g/cm^3 (or g/mL). When sugar (or another solute) dissolves in a liquid such as water, the volume of the solution does not differ significantly from the pure liquid. However, the mass of the solution increases. How will this affect the density of the solution?

Making Predictions

After reading *Gathering Evidence*, prepare a data table for your results. Develop several scientific questions that you can answer in this investigation.

Gathering Evidence

Overview of the Investigation

1. Prepare a hydrometer using a thin-stem plastic pipet.

2. Calibrate your hydrometer using sugar solutions of known concentration.

3. Use your data to prepare a calibration curve showing how hydrometer depth changes with sugar concentration.

4. Use your hydrometer to measure the density of several beverages.

5. Use your graph to determine the sugar concentration in your samples.

Part I. Constructing a Hydrometer

You will construct a hydrometer as shown in Figure 29.1.

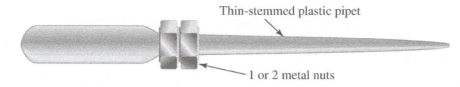

Thin-stemmed plastic pipet

1 or 2 metal nuts

Figure 29.1. A simple hydrometer.

1. Collect a thin-stemmed plastic pipet and one or two metal nuts of the type to fit on machine screws.

2. Slide the nut(s) over the stem of the pipet as shown in the figure.

3. Fill the pipet approximately half-full with water.

4. Place the pipet, bulb end down, into a 50-mL graduated cylinder filled nearly to the top with room temperature water. Adjust the amount of water in the pipet so that it floats with the bulb *near* the bottom (but not touching the bottom) and with only a short length of the stem sticking out of the water (Figure 29.2). If it floats too high or too low, adjust the height by cautiously adding or removing some water from the bulb of the pipette. When you have the pipette floating correctly, it is ready to use as a hydrometer.

To determine the relative density of a liquid, and hence the amount of dissolved solids (mostly sugar) in soft drinks or fruit juices, float the hydrometer in the beverage and measure the length of the stem that sticks out of the liquid. Before proceeding, discuss with your partner or lab group what you expect to happen when the hydrometer is placed into a beverage that contains sugar. Will more or less of the stem protrude from the surface of the liquid?

Part II. Calibrating the Hydrometer

The hydrometer you have constructed cannot measure the absolute concentration of sugar in a solution, but it measures the relative amount. To quantitatively estimate the amount of sugar in various beverages, you need to calibrate

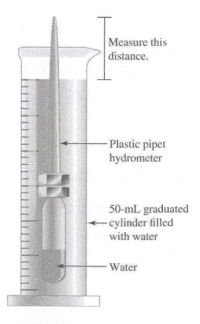

Measure this distance.

Plastic pipet hydrometer

50-mL graduated cylinder filled with water

Water

Figure 29.2. The hydrometer assembly.

the hydrometer you have constructed. In the lab, you will have beakers, graduated cylinders, stirrers, and sugar cubes available to prepare reference solutions, and rulers to measure the length of the pipet stem protruding from each solution. In your lab group, come up with a procedure for calibrating your hydrometer. Consider the following questions: How will you prepare the reference solutions? How many solutions should you make? How many sugar cubes will you put in each solution?

Your instructor should check your procedure before you do the calibration. Record your results in a table, and then plot the data on a graph (either by hand or on a computer). What should the axes of the graph be? Can your data points be connected by a straight line? Do you need to repeat any of the measurements? What sources of error might there be?

Part III. Testing Beverages

Using your calibrated hydrometer and the beverages provided by your instructor, answer the questions you developed in the *Making Predictions* section. Devise an investigation plan that will allow you to answer your questions.

The soft drinks need to be decarbonated before measuring the sugar content. Follow your instructor's guidance to do this. If the hydrometer is placed in a carbonated drink, bubbles sticking to the hydrometer will add buoyancy and the reading will be incorrect.

To avoid contaminating the samples, always rinse the hydrometer with pure water and blot it gently with a tissue before placing it in a new solution. Use the graph you generated in Part II to relate the height of your hydrometer stem to the concentration of sugar in the beverage. Record your data in a table, and share your results with the rest of your class. How do the results on the same beverage compare?

STOP! Do not consume any food or beverages that have been opened or used in the laboratory.

Optional Extension

If your instructor permits, you may take your hydrometer with you (being careful not to squeeze out any of the water) and use it to test other drinks at home. It is very important that you not drink any samples in which the hydrometer has been immersed. Measurements should be done on small samples and then discarded.

Analyzing Evidence

Use your calibration curve to determine the sugar content for each of the beverages you measured. Use your results to answer the questions you developed in *Making Predictions*.

Interpreting Evidence

1. Which beverages have the highest sugar content and which have the lowest? Did any of your results surprise you?

2. Do the results for any "sugar-free" beverages tested support this claim?

3. How does the sugar content of fruit juice compare to that of soft drinks?

4. Describe how you could modify your hydrometer to increase its accuracy.

Making Claims

What can you claim about the sugar content of beverages?

Reflecting on the Investigation

1. When cans of regular and diet soda are placed in a large container of ice water, some of the cans float while others sink. Which float? Which sink? Explain.

2. Ethanol (an alcohol) is found in beer and wine. Look up the density of ethanol and speculate how the density of a mixture of water and ethanol will change with increasing concentration of ethanol. Devise a method to determine the concentration of ethanol in alcoholic beverages using a hydrometer. How might you modify your hydrometer to make it suitable for measuring ethanol content of alcoholic beverages?

3. Ethylene glycol is a commonly used automobile radiator coolant. Most automobile coolant systems contain ethylene glycol mixed with water. Look up the density of ethylene glycol and devise a procedure for measuring the concentration of ethylene glycol in an automobile cooling system.

4. Find a bottle or can of regular soda and look at the label. What is the serving size? Is it the same as the size of the container? How many carbohydrates (in g) are there per serving? How many in the whole container? Given that one teaspoon of sugar (one sugar cube) has a mass of approximately 4 g, how many teaspoons of sugar would you consume if you drank the entire container of soda?

5. Discuss the environmental and health benefits of following a diet that contains foods low in added sugar.

6. Explain how the sugar present in fruit juice differs in chemical structure and origin from the sugar you used to make your reference solution and from that found in soft drinks. How do you think this affects your data?

7. In the United States, added sugar in processed foods is mostly in the form of high-fructose corn syrup (HFCS), not sucrose. There has been some controversy over the use of HFCS in foods, and many labels now claim that products are free of HFCS. Look into both sides of the issue and discuss the environmental and health impacts of HFCS vs. sucrose. Do you feel comfortable consuming foods containing HFCS? What about sucrose?

NOTES

Measuring Salt in Food

Asking Questions

- Why is salt added to food?
- What is the recommended daily allowance for salt consumption?
- Why should we be aware of consuming too much salt?
- What types of foods contain the most salt?

Preparing to Investigate

A moderate amount of salt (sodium chloride, $NaCl$) is essential in our diet for good health, but consumption of large quantities can have many adverse health effects. Large quantities of salt are commonly added to many processed foods and are often masked by other flavoring agents. In this investigation, you will measure the salt content in typical servings of some packaged soups and in the liquid from a jar of pickles.

The easiest way to measure salt content in foods is to measure the concentration of chloride ions by means of a **titration**. For this investigation, you will add a solution of silver nitrate ($AgNO_3$) to a measured quantity of liquid sample until just enough has been added to react exactly with all of the chloride ions present in the sample. The reaction forms silver chloride ($AgCl$), an insoluble white solid.

$$AgNO_3(aq) + Cl^-(aq) \rightarrow AgCl(s) + NO_3^-(aq)$$

If the amount of silver nitrate added to a sample is known, then the chloride ion concentration can be calculated because we know that 1 mole of Ag^+ reacts with 1 mole of Cl^-.

To better see the end point of the titration, you will add an **indicator**, sodium chromate (Na_2CrO_4). Chromate ion (CrO_4^{2-}) reacts with Ag^+ to form silver chromate (Ag_2CrO_4), an insoluble red precipitate. As you add $AgNO_3$ to your sample, Ag^+ will react with Cl^- to form $AgCl$. Once all of the chloride ions have reacted, the additional silver ions will react with CrO_4^{2-} and form a red precipitate. The presence of the red color signals the end of the titration.

For this investigation, you will assume that all of the chloride that you measure comes from salt ($NaCl$). This assumption is not always a good one because some foods may have chloride ions present that are associated with other metal ions such as calcium, potassium, or magnesium. However, for processed or preserved foods such as bouillon cubes, canned soup, or pickles, the majority of the chloride ions do indeed come from added salt.

Making Predictions

After reading *Gathering Evidence,* prepare a data table. Read the labels on your assigned samples, and rank them according to salt content.

Gathering Evidence

Overview of the Investigation

1. Devise a procedure for titrating chloride in foods.

2. Perform the titration on dissolved bouillon cubes, soup, and pickle liquid.

3. Calculate the salt concentrations in the samples.

 CAUTION! Be careful not to spill the silver nitrate or sodium chromate solutions on your skin. Silver nitrate will stain your skin black; it is harmless and wears off in a few days, but contact should still be avoided. More seriously, sodium chromate is a **carcinogen** and contact with it must be avoided.

Part I. Developing Your Procedure

Titration techniques are described in the *Laboratory Methods* section. The procedure for the titration of chloride with silver nitrate is described in detail in Investigation 18. Read the background and procedure in Investigation 18, and adapt that procedure for use with foods. Write down your procedure and have your instructor check it before proceeding.

Part II. Performing Titrations on Food Samples

A. Bouillon and other soups

1. Obtain a bouillon cube and dissolve it in 1 cup, or 227 mL, of water. The cube will dissolve faster if you crush it, heat the water, and stir. Be sure to stir the solution until the cube is completely dissolved. *Note*: One bouillon cube will provide enough broth for several student groups, so you can share your sample.

2. Following the procedure you developed in Part I, carry out the titration starting with 20 drops of broth. Record the number of drops of silver nitrate required to reach the end point of the titration. Do at least three titrations with the broth. If you are uncertain about the accuracy of any of your titrations, perform one or two more.

3. If other soups are available, do at least three titrations on those samples. Record the number of drops used. Rinse your sample pipet thoroughly before each new sample.

B. Pickle liquid

1. Pickle liquid is typically very high in salt, so you should use a smaller volume of the sample than you did for the bouillon. Start each titration by putting 5 drops of pickle liquid into each well. Add 15 drops of water to each well to dilute the pickle liquid.

2. Perform your titration at least three times with the pickle liquid. If the first titration uses too many or too few drops of silver nitrate, adjust the dilution of pickle liquid accordingly for subsequent trials.

Clean up

Empty your solutions into the designated waste container, and clean your equipment thoroughly.

Analyzing Evidence

A detailed explanation of the required calculations is given in Investigation 18. After reading the appropriate section in that investigation, perform the required calculations using your data from this investigation.

1. First, look carefully at your titration results for each sample. Do the results all agree with each other reasonably well? If one of the results seems out of line compared to the others, it is reasonable to omit that result. Simply draw a line through that row of your data table (do not scribble it out!) and write a note saying why you are omitting that result.

2. Calculate the average number of drops of silver nitrate solution that were required to titrate each sample, and record this value on your data table.

3. You can calculate the molarity of the chloride ion in the sample using Equation 1 below, which was derived in Investigation 18. Do this calculation and record the molarity of chloride for each sample you analyzed.

$$\text{molarity of Cl}^- = \text{molarity of AgNO}_3 \times \frac{\text{drops of AgNO}_3}{\text{drops of sample}} \tag{1}$$

4. We aren't really interested in the *molarity* of chloride ions in the soups or pickles, but how much salt and, more particularly, how much *sodium* is present in a serving. Because the formula for salt is $NaCl$, the molarity of sodium ion (Na^+) is equal to the molarity of chloride ion (Cl^-).

The mass of one mole of sodium is 23 grams, or 23,000 milligrams. Thus, multiplying the molarity of sodium ions by the molar mass gives you the mass, in milligrams, of sodium per 1 L of solution. For example, suppose you measure the concentration of chloride to be 0.20 M. The sodium concentration will also be 0.20 M, so

$$\frac{0.20 \text{ moles sodium}}{1 \text{ L of soup}} \times \frac{23{,}000 \text{ mg sodium}}{1 \text{ mole sodium}} = \frac{4600 \text{ mg sodium}}{1 \text{ L of soup}}$$

Multiplying that value by the size of one serving provides the amount of sodium per serving. Supposing a serving size of 1 cup, or 227 mL, and given that 1 L = 1000 mL,

$$\frac{4600 \text{ mg sodium}}{1000 \text{ mL soup}} \times \frac{227 \text{ mL soup}}{1 \text{ serving of soup}} = 1044 \text{ mg sodium per serving}$$

Perform this calculation for all servings of bouillon and soup that you analyzed. For the pickle liquid, calculate the milligrams of sodium in 100 mL of liquid.

Interpreting Evidence

1. Look up the recommended daily intake for sodium. If you consumed 1 cup (227 mL) of the bouillon or other soup that you analyzed, what percentage of the recommended daily intake of salt would you have consumed?

2. If a pickle is 90% fluid, how much sodium would be found in a 90 gram pickle of the type you estimated? What percentage of your recommended daily intake of salt would be supplied by one pickle?

Making Claims

What can you claim about salt in foods?

Reflecting on the Investigation

1. Design a procedure to determine how much salt is contained in a small bag of potato chips or in a small bag of salted peanuts. In designing the investigation consider the following:

    a. How would you choose a representative sample of the chips or nuts?

    b. How would you remove the salt from the food sample?

    c. If you washed the salt off the chips or nuts, would it matter how much water you used?

    d. With solid samples, you may wish to use different units than you used for liquid samples. How would you report your results?

2. Packaged lunch meats also contain large quantities of salt. How could you estimate the amount of salt in these products?

Measuring Vitamin C in Juice and Tablets

Asking Questions

- How much vitamin C is in juice and vitamin tablets?
- How can we measure the amount of vitamin C?
- Why is vitamin C important to our diet?
- What are macronutrients and micronutrients? Which of these terms applies to vitamin C?

Preparing to Investigate

Vitamin C, or ascorbic acid ($C_6H_8O_6$), is one of the essential vitamins required for good health. It is chemically similar to the simple sugar glucose ($C_6H_{12}O_6$), which is plentiful in our bodies. Most animals possess an enzyme needed for making ascorbic acid from glucose, but humans and a few other species lack that enzyme. Therefore, we must secure ascorbic acid directly from foods that we consume. One function of ascorbic acid is the prevention of scurvy; a minute amount of vitamin C (10 mg per day) is adequate to ward off this condition; however, it also plays several other important roles in human health. Because it has a strong tendency to transfer electrons to other chemical substances, vitamin C is classified as a reducing agent or antioxidant. It prevents potentially harmful reactions that involve the transfer of electrons.

To maintain good health, the recommended daily intake (RDI) for vitamin C is 60 mg. Foods like fruits and vegetables with high water content often contain large amounts of vitamin C. Ascorbic acid is very soluble in water, so if high doses are ingested much of it is excreted rapidly in urine.

In this investigation, you will analyze fruit juice and vitamin C tablets to determine how much vitamin C is contained in each. You will use the **titration** method of analysis, which is described in detail in the *Laboratory Methods* section of this lab manual. The chemistry of the vitamin C titration is based on its tendency to lose electrons. You will use iodine (I_2) as the electron acceptor to titrate the ascorbic acid. The chemical reaction is:

$$C_6H_8O_6 + I_2 \rightarrow C_6H_6O_6 + 2 I^- + 2 H^+$$

Since it is difficult to prepare iodine solutions with a specific concentr~~ation~~ first ~~col~~ored, titrate your iodine solution using a reference sample of ascorbic acid. Be~~cause~~

the endpoint of the titration is signaled by the appearance of the color of unreacted iodine, which happens when all of the ascorbic acid has reacted. The color is made more intense by adding a drop of starch solution, which forms a deep blue color in the presence of iodine.

Making Predictions

After reading *Gathering Evidence*, prepare a data table where you will record your results. Develop several scientific questions to answer by analyzing the vitamin C content of the samples available in your lab. Looking at the labels on the samples, predict the relative amounts of vitamin C in each of the samples you will titrate. Which samples should have the most vitamin C and which should have the least?

Gathering Evidence

Overview of the Investigation

1. Prepare a solution with a vitamin C tablet in a volumetric flask.

2. Titrate the ascorbic acid reference solution with iodine solution.

3. Titrate the vitamin C tablet solution and any juice samples with the iodine solution.

4. Calculate the amount of vitamin C in your samples.

Part I. Preparing the Vitamin C Tablet Solution

1. Weigh a vitamin C tablet and record its mass to the nearest milligram (0.001 g).

2. Place the tablet in a clean volumetric flask. Fill the flask about half full with water. Shake and swirl the flask until the tablet is broken down. It may be helpful to gently crush the tablet with a glass stirring rod. Label the flask and allow it to settle.

Part II. Titrating an Ascorbic Acid Reference Solution

Before proceeding, review the section about titration in the *Laboratory Methods* section. You will be titrating a solution of ascorbic acid of known concentration to determine the concentration of your iodine solution. The procedure here can also be followed to titrate your samples.

1. Use a graduated-stem plastic pipet to add exactly 1 mL of the ascorbic acid reference solution into four wells in a clean, dry wellplate. Place the wellplate on a white surface to make it easier to observe color changes.

   drop of starch solution to each well.

3. Obtain a small supply of iodine solution in a clean container. Use a pipet to add one drop of iodine at a time to your solution. Count the drops and stir between additions. You'll notice a momentary blue color that disappears with stirring but lingers longer as more iodine is added. The endpoint occurs when one drop of iodine gives a pale blue color that does not disappear with stirring. Record the number of drops that lead to this color change.

Part III. Titrating Vitamin C Tablet Solution

By now, the vitamin C tablet in the volumetric flask should be dissolved, except for the insoluble filler material, which is usually starch. Add deionized water to the flask up to the bottom of the neck. Then use a pipet to add water *one drop at a time* until the bottom of the curved meniscus of the liquid surface just touches the etched line on the neck of the flask (Figure 31.1). Cap the flask securely and mix the contents of the flask by repeatedly turning the flask upside down and swirling.

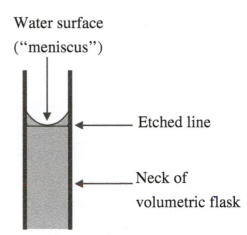

Figure 31.1. How to use a volumetric flask.

Rinse the graduated-stem pipet you used for the reference solution with deionized water (fill and empty it several times), and then rinse with solution from the volumetric flask in the same manner. Do the titration in four wells as described in Part II and record your data.

Part IV. Titrating Juice Samples

Use the same procedure described above to titrate one or more samples of juice. You can use bottled juice, or you can extract and analyze the liquid from fruits and vegetables that are high in vitamin C (such as peppers and broccoli). You should do at least four titrations for each sample. Be sure to thoroughly rinse your graduated-stem pipet between samples with deionized water and with the new sample to be analyzed before carefully adding 1 mL of sample to each well in the wellplate. Record your data.

CAUTION! Never consume food or beverages used in the laboratory.

Analyzing Evidence

Look at your data and decide whether the data for each set of titrations shows consistency. If any individual result should be excluded because of an error, make a note beside that value and omit it from your calculations. Then, calculate the average number of drops of iodine used for each set of titrations.

First, you will need to calculate the *calibration factor* for this analysis to give you a way of correlating the amount of ascorbic acid in your samples to the amount contained in the reference solution. Divide the concentration of the reference solution, in mg/mL, by the number of drops of iodine used in the titration, as shown below. You will then know how many milligrams of ascorbic acid are titrated by one drop of iodine.

$$\frac{\text{mg ascorbic acid}}{\text{1 mL reference solution}} \times \frac{\text{1 mL reference solution}}{\text{drops of iodine used}} = \frac{\text{mg ascorbic acid}}{\text{1 drop of iodine}}$$

For the analysis of your samples, multiply your calibration factor by the average number of drops of iodine used for each titration.

$$\frac{\text{drops of iodine used}}{\text{1 mL of sample}} \times \frac{\text{mg ascorbic acid}}{\text{1 drop of iodine}} = \frac{\text{mg ascorbic acid}}{\text{1 mL of sample}}$$

This allows you to calculate the milligrams of ascorbic acid present in 1 mL of sample. For the tablet, multiply the answer by 100 to find the total milligrams of ascorbic acid in the 100-mL volumetric flask. This is the same as the milligrams of ascorbic acid in one tablet.

Interpreting Evidence

1. Answer the scientific questions you posed in the *Making Predictions* section.

2. What fraction of the mass of the vitamin tablet was ascorbic acid?

3. How closely did your analysis of the vitamin tablet match the label on the bottle?

4. If a standard serving of juice is 8 ounces and 1 fluid ounce = 30 mL, calculate how much vitamin C would be obtained from an 8-ounce serving of the juice you analyzed. Does this match the amount cited on the label? How many servings would you need to consume to meet the RDI for vitamin C?

Making Claims

What can you claim about the amount of vitamin C in various foods and juices?

Reflecting on the Investigation

1. Is vitamin C a water-soluble or fat-soluble vitamin? How does this relate to the procedure we used for analysis?

2. The RDI cited in the introduction is the recommendation to the general population by the U.S. Food and Drug Administration for the minimum amount that will allow most people to maintain health. In 2004, the Institute of Medicine (a branch of the National Academies) published more detailed Dietary Reference Intake (DRI) values, which take into account age, sex, and activity levels of individuals when recommending nutrient intake levels.

    a. Locate the tables and find your DRI for vitamin C. How much vitamin C should you take in each day?

    b. How much juice or other foods that you analyzed would you need to ingest to meet your DRI?

    c. Is DRI a better suggestion of nutrient intake than RDI? Why or why not?

    d. What foods are rich in vitamin C? Do you think that your diet provides you with enough vitamin C to meet your DRI?

NOTES

Measuring Iron in Cereal

Asking Questions

- What form of iron is added to fortified breakfast cereal?

- Is cereal a compound or a mixture?

- What is one way that materials can be extracted from mixtures?

- Why is iron biologically important?

- How is the iron made available for use in our bodies?

Preparing to Investigate

In this investigation, you will look at a biologically important compound and follow it from a nutritional source through a transformation that makes it useful in our bodies. You will extract elemental iron from breakfast cereal, dissolve it in "stomach acid," and then confirm that it has been converted to a compound that includes the Fe^{2+} ion.

Iron is involved in many biological processes. The most important of the functions of iron is to bind to oxygen atoms and transport them in our bloodstream. Oxygen provides energy for all the tissues and organs in our body, so iron is essential for life. Since our bodies cannot produce iron, we need to take it in through our food. The iron in foods exists in different forms. The most easily absorbed iron is iron that is associated with the protein hemoglobin. Beef liver and mollusks contain high amounts of this form of iron. Other sources of iron include meats, fish, lentils, spinach, and iron-fortified foods.

Cereals are one of the most popular fortified foods. Many vitamins and minerals are added to these grain products to improve their nutritional value. Iron is added to cereals in its elemental form as small metal filings. While it is not as available to our bodies as the iron in beef liver, there are chemical processes in our bodies that can transform it to improve its usefulness.

Making Predictions

- What property can you use to extract iron from a mixture? Is this a physical or chemical process?

- Most metals react with acidic solutions. What do you think happens as iron filings pass through our digestive system? Explain your prediction.

- After reading *Gathering Evidence*, prepare a data sheet to record your observations, measurements, calculations, and UV-visible spectral data.

Gathering Evidence

Overview of the Investigation

1. Observe the interactions between flakes of cereal on the surface of water and a magnet.
2. Grind cereal into a fine powder. Stir cereal and warm water in a large beaker.
3. Stir using a strong magnet in a small zipper bag.
4. Decant and rinse the precipitate with water and ethanol.
5. Determine how much iron has been extracted and compare to the expected value.
6. Dissolve iron filings in HCl with heat.
7. Test for Fe^{2+} formation by reacting the acidic solutions with o-phenanthroline.
8. Confirm with UV-visible spectrophotometry.

Part I. Isolating Iron from Cereal

1. Fill a Petri dish with water so that the water is over the brim but not spilling. Put a flake or two of a cereal with a high iron content in the center of the dish.
2. Put a strong magnet near the flake and move the magnet slowly, staying parallel to the surface of the water. Record your observations.
3. Use a mortar and pestle to grind approximately 45 g of the cereal into a fine powder. Record the exact amount that you use.
4. Mix the ground cereal and 375 mL of warm water in a 600-mL beaker.
5. Stir the mixture with a plastic spoon or glass stir rod for 20–30 minutes. The cereal should soften and form a slurry.
6. Put a strong magnet in a small plastic zipper bag and seal it. Use a balance to measure the mass of the magnet and bag. Record the mass.
7. Add the magnet to the cereal slurry by tipping the beaker and sliding the magnet down the side of the beaker.
8. Use the magnetic stirrer on a stir plate to stir the slurry with the magnet for 10 minutes. Adjust the speed so that the stirring forms a vortex without splashing.

9. Remove the beaker from the stir plate and decant as much of the cereal as possible. Add enough water to rinse the sides of the beaker and cover the plastic bag.

10. Swirl and decant the water. Repeat with ethanol.

11. Place the magnet (still inside the plastic bag) in a weighing boat and allow it to dry.

12. Place the dry magnet and bag on a piece of weighing paper and measure the mass.

13. In a weighing boat, open the bag (with the zipper on top) and carefully flip the bag inside out, while pulling the magnet upward. The filings should fall into the weighing boat.

14. Use weighing paper to measure the mass of the filings. Record the mass.

Part II. Converting the Iron Metal

1. Transfer approximately half of the iron filings to a 125-mL Erlenmeyer flask and add 50 mL of 0.1-M HCl and a stir bar.

2. Put a watch glass on top of the flask and heat the solution to approximately 37 °C for as long as possible (four hours is ideal but probably impractical in a teaching laboratory).

3. Transfer the other half of the iron filings to a different 125-mL Erlenmeyer flask and add 50 mL of 1-M HCl and a stir bar.

CAUTION! HCl is corrosive and can harm skin and eyes. Wear goggles and gloves and notify your instructor of any spills. If any acid contacts your skin, rinse immediately.

4. Put a watch glass on top of the flask and heat the solution to approximately 37 °C for 20 minutes.

5. Remove the solutions from the heat and allow them to cool to room temperature.

6. Record your observations about the state of the iron filings in the two flasks.

(Optional) Part III. Confirming the Presence of Fe^{2+} Ions

1. Transfer 15 drops of one of the acidic solutions to a labeled test tube. Repeat for the other solution.

2. Add 2 mL of distilled water to each test tube.

3. Collect UV-visible spectra for these solutions (see the Spectroscopy section in the *Laboratory Methods* section of this lab manual). Follow the procedure outlined by your instructor.

4. Add one drop of o-phenanthroline solution to each test tube. Record observations.

5. Collect UV-visible spectra for these solutions.

Clean up

Discard or recycle the contents of your beakers and flasks as directed by your instructor.

Analyzing Evidence

1. Using the masses of the magnet in the plastic bag before and after being in the cereal slurry, calculate the amount of iron that was extracted from the cereal.

2. Using the measurement of the mass of the iron filings that came off of the plastic bag, determine the amount of iron that was extracted from the cereal.

Interpreting Evidence

1. What did you notice when you put the magnet next to the cereal flake? What property allowed the flake to move with the magnet? What form must the iron be in to display that property?

2. How do the measurements of iron extraction compare to each other? Why might they be different?

3. Compare the measurements of iron extraction to the expected amount according to the cereal box label. Make sure you do your calculations based on the actual amount of cereal that you used. Did you extract the expected amount? If it is different, what could have contributed to this difference?

4. How are the concentrations of acid and temperatures that you used to convert the iron filings to Fe^{2+} biologically relevant?

Making Claims

What can you claim about the iron in fortified cereal?

Reflecting on the Investigation

1. If you collected UV-visible spectra of the iron(II) phenanthroline complexes, compare them with known spectra. Do your spectra confirm that Fe^{2+} ions were formed when the iron filings were dissolved in acid?

2. Go back through the investigation and identify any physical changes and chemical changes that occurred. How do you know which change falls under each category?

Analysis of Vinegar

Asking Questions

- How much acid is in vinegar?
- What is the acidic compound in vinegar?
- How can you measure acidity?

Preparing to Investigate

As described in Chapter 8 of *Chemistry in Context*, solutions of **acids** in water contain more H^+ ions than OH^-, while solutions of **bases** contain more OH^- ions than H^+ ions. Vinegar is a dilute aqueous solution of acetic acid ($HC_2H_3O_2$). It results from the oxidation of ethanol, which is produced by fermenting sugars, as described in Chapter 10. Acidic solutions containing vinegar, and usually salt, are commonly used for preserving and pickling foods.

A characteristic property of acids and bases is that they react with each other. Thus, acetic acid in vinegar will react with sodium hydroxide, a base, to produce water and sodium acetate.

acetic acid	+	sodium hydroxide	→	water	+	sodium acetate
$HC_2H_3O_2$	+	NaOH	→	H_2O	+	$NaC_2H_3O_2$

This type of reaction is the basis for the **titration** method of acid analysis. This method is described in detail in the introductory *Laboratory Methods* section of this lab manual. When using titration to analyze acids, a known quantity of acid solution is measured. Then, a solution of base is added slowly until just enough has been added to react with all of the acid. If the concentration of base is known and the volume of added base is carefully measured, you can calculate how much acid must have been present. Finally, if the volume of the acid solution is known, you can calculate the concentration of the acid. Concentrations are expressed in **molarity (M)**, which is defined as the moles of a substance in exactly one liter of solution.

In a titration, an **indicator** is added so that a color change occurs to show when a reaction has taken place. The indicator used in this investigation is *phenolphthalein*, which is colorless in acid and pink in basic solutions. The endpoint of your titration will occur when one drop of base solution changes the solution from colorless to pink. Indicators are intensely colored so only one drop needs to be added to your titration set up.

Making Predictions

- Look up and draw the structure of acetic acid, showing how the atoms are bonded together. What functional group does it contain? See Chapter 9 in *Chemistry in Context* for help. Speculate on why acetic acid is acidic. Which hydrogen makes the H^+? How does it react with NaOH?

- After reading *Gathering Evidence*, prepare a data table to record your titration results.

Gathering Evidence

Overview of the Investigation

1. Obtain a 24-well wellplate, four plastic transfer pipets, and necessary solutions.

2. Practice dispensing drops evenly.

3. Practice titrating vinegar with sodium hydroxide solution.

4. Titrate samples of vinegar to determine their acetic acid content.

5. Calculate the molarity and percent acidity of the acetic acid in the vinegar.

Part I. Setting Up the Titration

1. Obtain your materials and label three pipets with vinegar, NaOH, and indicator. Fill the vinegar and NaOH pipets with the appropriate solution, and partially fill the "indicator" pipet with indicator.

     CAUTION! Sodium hydroxide solution is corrosive and can cause serious skin and eye damage. Therefore you MUST wear eye protection at all times and avoid getting any of the solution on your skin. In case of skin contact, rinse immediately with plenty of water and notify your instructor.

2. Follow the instructions given in the *Laboratory Methods* section to practice using a pipet and conduct a sample titration. It will help to place your wellplate on a white piece of paper to make the color change easier to see.

Part II. Measuring the Acetic Acid Content of Vinegar

Remember when titrating, your goal is to catch the point where *one drop* of NaOH produces the first permanent color change. Record your results on your data table.

1. Carefully add <u>20 drops</u> of vinegar (count them) to each of six wells. Make sure that the drops fall directly to the bottom of the well and are not "trapped" along the side. If you make a mistake, start again in another well. Record the position numbers for the wells you are using.

2. Add <u>1 drop</u> of indicator to each of the six wells containing 20 drops of vinegar.

3. Follow the instructions in *Laboratory Methods* to titrate each well to the endpoint where one drop of NaOH produces a permanent magenta color. Record the number of drops used. If you make a mistake, you may skip a trial and go on to the next one.

Note: If you miss the endpoint, lose count of drops, or are uncertain of the result for any other reason, simply make a note in your data table about what you think went wrong, draw a line through that row, and do another trial. All scientists make mistakes when doing lab work. You'll have plenty of data to record good results.

Optional extension: Analyze another kind of vinegar. First, rinse the vinegar pipet with the new sample twice by filling the pipet and emptying it into the appropriate waste container. Use a new set of wells in the wellplate for the new sample.

Clean up

Your instructor will specify where to dispose of the contents of the wellplates and whether to wash or dispose of the plastic pipets. Wash the wellplate thoroughly and allow it to dry.

Analyzing Evidence

Recall that you will be using the known concentration of NaOH and the volumes of NaOH and vinegar that you used to calculate the concentration of acetic acid in your vinegar solution. This is the balanced equation for the reaction of acetic acid with sodium hydroxide:

$$HC_2H_3O_2 + \quad NaOH \rightarrow \quad H_2O + NaC_2H_3O_2$$

This equation shows that *1 mole of acetic acid reacts with 1 mole of sodium hydroxide*. Thus, for the titration of vinegar, the number of moles of NaOH added from the pipet is exactly equal to the number of moles of acetic acid in the vinegar in the well. Expressed mathematically, we can say that

moles of NaOH added from the pipet = moles of $HC_2H_3O_2$ in the well

The number of moles of NaOH in any sample of NaOH solution can be calculated from the molarity of the NaOH (expressed as moles per liter, or mol/L) and the volume (in liters), using the following equation:

$$\text{moles of NaOH used} = \text{liters of NaOH solution used} \times \frac{\text{moles of NaOH}}{1 \text{ liter of NaOH solution}}$$

A similar equation can be written for the moles of acetic acid in the vinegar sample.

$$\text{moles of acetic acid used} = \text{liters of vinegar} \times \frac{\text{moles of acetic acid}}{1 \text{ liter of vinegar}}$$

So if we rearrange some of the above equations, we can conclude:

$$\text{liters of NaOH} \times \frac{\text{moles of NaOH}}{\text{liters of NaOH}} = \text{liters of vinegar} \times \frac{\text{moles of acetic acid}}{\text{liter of vinegar}}$$

or: $$\text{liters of NaOH} \times \text{molarity of NaOH} = \text{liters of vinegar} \times \text{molarity of acetic acid}$$

So:

$$\text{molarity of acetic acid} = \text{molarity of NaOH} \times \frac{\text{liters of NaOH}}{\text{liter of vinegar}}$$

There is one small problem that prohibits you from using this equation –you don't know the volumes of your solutions in liters. You only know the number of drops of each solution that you used. If we assume that all of the drops from both pipets have the same volume, then the ratio of volumes expressed as drops should be the same as the ratio of volumes expressed as liters.

$$\frac{\text{liters of NaOH}}{\text{liter of vinegar}} = \frac{\text{drops of NaOH}}{\text{drops of vinegar}}$$

So to calculate the molarity of acetic acid, we can derive this relationship:

$$\text{molarity of acetic acid} = \text{molarity of NaOH} \times \frac{\text{drops of NaOH}}{\text{drops of vinegar}}$$

If you look at the label on the vinegar you used, you may see the amount of acetic acid reported not as molarity, but as *percent acidity*, which is defined as the number of grams of acetic acid per 100 mL of vinegar solution. One mole of acetic acid weighs 60 grams, and 100 mL is 1/10 of a liter; therefore the percent acidity can be calculated this way:

$$\text{percent acidity} = \frac{\text{moles of acetic acid}}{\text{liter of vinegar}} \times \frac{60 \text{ grams of acetic acid}}{1 \text{ mole of acetic acid}} \times \frac{1}{10}$$

Calculations

1. Look carefully at the titration results recorded in your data table. Does any one result seem to differ greatly from the others? If so, a mistake may have been made, and it is legitimate to omit that result. Simply write a comment on your table next to that result saying that you are leaving it out.

2. Use all of the results that you think are valid to calculate the average number of drops of NaOH and record this on the data sheet.

3. Next, calculate the molarity of acetic acid in the vinegar, using the average number of drops of NaOH.

4. Lastly, calculate the percent acidity of your vinegar sample.

Interpreting Evidence

1. How does your calculated molarity compare to other acidic substances? Is it more or less acidic than stomach acid (about 0.16 M HCl)? How does it compare to acid rain (see Chapter 8 in *Chemistry in Context*)?

2. Why do you suppose that the acetic acid content in vinegar is reported as percent acidity rather than as molarity?

Making Claims

What can you claim about the acidity of vinegar based on your results from this investigation?

Reflecting on the Investigation

1. What is the advantage of doing several trials, especially since you are doing the exact same thing each time?

2. Suppose that during the titration, the addition of NaOH changed the color of the solution to a dark pink instead of a light pink. How would this change the results? Would your calculated molarity be higher or lower than the actual molarity? Explain.

3. Could you have done the titration "backwards," starting with 20 drops of NaOH, adding the phenolphthalein indicator, and then adding the vinegar one drop at a time? Explain. What would you observe? Would the endpoint be easier or harder to see than with the titration method you used?

4. Can you think of ways to improve this method of analysis? What changes might be possible to give you more accurate results?

5. Titration can be used to determine the acid content of any food or drink.

    a. List three foods or drinks that you believe contain acid. *Hint:* Acidic foods often taste tart.

    b. Why would titrations of highly colored substances require some modification to the procedure?

Identifying Analgesic Drugs with TLC

Asking Questions

- What is an analgesic?
- What compounds are often in an analgesic tablet?
- What substances other than the active ingredients are in an analgesic tablet?
- What properties of a molecule can allow it to be separated from other molecules?
- How can you identify the compounds present in a tablet?

Preparing to Investigate

Over-the-counter analgesic preparations are widely sold and used worldwide. Aspirin (acetylsalicylic acid) is a common analgesic that is used for pain and fever relief. Some people take a low dose of aspirin as a preventative medicine for heart attacks and strokes due to its anti-clotting activity. Other common non-prescription analgesic drugs include acetaminophen, ibuprofen, and naproxen. Caffeine is sometimes added to analgesic tablets to overcome drowsiness and increase the pain-relieving properties of the medicine. Buffered aspirin contains a base such as magnesium hydroxide or calcium carbonate to neutralize the acidic compound and help prevent stomach irritation. Analgesic tablets also contain binders, such as starch and sugars, and smooth coatings to ease swallowing.

Sometimes, forensic scientists need to identify an unknown substance that is in some way related to a crime, and in this investigation, you will use thin-layer chromatography (TLC) to identify the analgesic compound(s) present in an over-the-counter painkiller and an unknown drug. TLC is described in detail in the *Laboratory Methods* section of this lab manual. It is one of the easiest chromatographic techniques and allows you to separate and identify compounds based on their differing affinity for the stationary phase (usually polar silica or alumina coated on a plastic or aluminum plate) and the mobile phase (less polar organic solvents).

Making Predictions

After reading *Gathering Evidence* and the section on TLC in *Laboratory Methods*, explain how TLC will be used to identify the components of an analgesic tablet. Look up the structures of the three most common analgesics—aspirin, acetaminophen, and ibuprofen—and caffeine. Identify structural differences that may contribute to different activity in TLC.

Gathering Evidence

Overview of the Investigation
1. Prepare a TLC developing chamber.
2. Spot known and unknown samples on a chromatographic plate.
3. Develop the chromatogram.
4. Calculate R_f values.
5. Identify components in an unknown sample.

Part I. Preparing the TLC Developing Chamber
1. Obtain a wide-mouth jar with a lid or a large beaker (400–600 mL) with a piece of plastic wrap or foil to cover it.
2. Prepare your developing chamber as shown in Figure 0.9 in the *Laboratory Methods* section of this lab manual, with a piece of filter paper on the wall of the chamber and about 5 mm of solvent mixture (containing 25 parts ethyl acetate, 1 part ethanol, and 1 part acetic acid) in the bottom. Cover the container and set it aside while preparing the chromatographic plate.

Part II. Preparing the Sample
1. Obtain an unknown analgesic tablet (or a portion of one) and record the sample number.
2. Use a mortar and pestle to crush the tablet to a fine powder.
3. Add 2–3 mL of methanol to the powder and stir. Allow the mixture to settle.

Part III. Preparing the Chromatographic Plate
1. Obtain your chromatographic plate. Be careful not to bend or twist the sheet, and handle it only by the edges.
2. Use a pencil (not a pen!) to prepare your plate (Figure 34.1), following the instructions in *Laboratory Methods*. Evenly space the spots and label each point with a penciled number or abbreviation so you know where to place each sample.

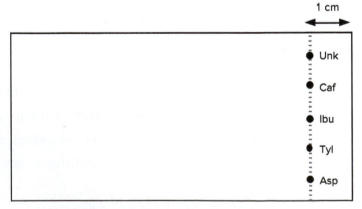

Figure 34.1. Spotting the TLC plate.

3. Before spotting your plate, you should practice putting *very small* spots of solution on a piece of scrap paper. To do this, dip the tip of a glass capillary tubes that has been drawn down to a very small opening into a solution. Very gently touch the tip to the paper for a brief moment. Be careful because the tubes are both sharp and fragile.

4. Use capillary tubes to spot your samples onto the TLC plate. The spots should be as small as possible to minimize tailing and overlapping when the chromatographic sheet is developed. If a more intense spot is desired, let the spot dry and re-spot in the same location.

5. Known solutions of aspirin, acetaminophen, ibuprofen, and caffeine will be available in the lab with a glass capillary in each. Use the capillary tubes to place small spots of the four solutions at their designated pencil marks. Your instructor may ask you to add another spot with a mixture of all four compounds. Use a clean glass capillary tube to spot a sample of the clear solution from your unknown tablet onto the chromatographic plate. Allow the dots' solvents to evaporate.

Part IV. Developing the TLC Plate

1. When the spots are dry, place the plate in the developing chamber. Check to be sure that the bottom edge (near the spots) is in the solvent, but that the spots are above the solvent. Also, be certain that the filter paper does not touch the TLC plate. Cover your chamber and watch carefully. The liquid will slowly move up the TLC sheet by capillary action.

2. When the front edge of the liquid has moved to 1 cm below the top of the plate, remove the plate from the TLC developing chamber. *Immediately*, while the sheet is still wet, draw a pencil line on the sheet to show the top edge of the liquid. Then, lay the sheet on a clean surface in a fume hood or other well-ventilated area and allow the solvent to evaporate until the sheet appears dry.

3. The spots are unlikely to be visible to the naked eye, but they should be easy to see when viewed under an ultraviolet (UV) lamp. Observe your plate under the UV lamp, and carefully outline any visible spots in pencil.

 CAUTION! UV radiation is harmful to your eyes. Never look directly into the UV lamp.

Analyzing Evidence

1. Calculate the R_f for your known samples and for each spot in your unknown sample. See the *Laboratory Methods* section if you need a reminder for how to do this. Report your results in a table.

2. Tape your TLC plate to a piece of paper and cover the whole plate with clear tape to protect it. Alternatively, draw your plate and dispose of it, as directed by your instructor.

Interpreting Evidence

1. Did the three pure analgesic compounds and caffeine each produce a single spot? Did the spots move different distances?

2. If you were given a known mixture of analgesic compounds to analyze, did the number of spots match the expected number of compounds? Did each spot move the same distance up the plate as the corresponding pure compound?

3. How many spots are on your TLC plate for your unknown sample? How many active compounds are present in your unknown sample? Can you identify the compounds based on their R_f values?

Making Claims

What can you claim about the identity of your unknown sample?

Reflecting on the Investigation

1. Why was it important to use a small amount of sample when spotting the plate?

2. Suggest possible advantages and disadvantages of using a longer (taller) TLC plate.

3. If two components have an identical R_f value, does this mean they necessarily have the same structure? Explain why or why not.

4. Do you expect that changing the solvent will change the R_f value for a given substance? Explain your reasoning.

5. The relative movement of components is controlled partially by the polarity of the molecules. The TLC sheet is coated with a highly polar substance, and the solvent mixture has a much lower polarity. From your chromatographic results, predict the relative polarities of aspirin, acetaminophen, ibuprofen, and caffeine. Explain your reasoning.

6. In an effort to identify an unknown, some students obtained a TLC plate (shown).

    a. How many compounds are in Sample A? In Sample B? In the unknown?

    b. A student concludes that the unknown is the same as Sample B because of the number of spots. Is this a valid conclusion? Explain.

    c. Another student concludes that the unknown is the same as Sample A because they both have spots with R$_f$ values of 0.3. Is this a valid conclusion? Explain.

    d. Propose a reasonable conclusion for the identity of the unknown. Explain your reasoning.

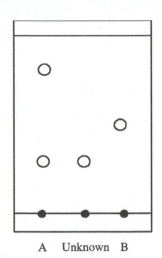

A   Unknown   B

NOTES

Drugs in the Environment

Asking Questions

- List some of the pharmaceutical and personal care products that you use each day.
- How might these products enter the environment?
- What are some consequences when these chemicals enter the environment?

Preparing to Investigate

Pharmaceuticals and personal care products (PPCPs) enhance or improve our quality of life and can enable us to live longer and more comfortably. However, due to improper disposal, incomplete metabolism, and other routes, PPCPs often end up in our waterways and soils (Figure 35.1). These compounds can have an adverse effect on the environment. Some compounds decompose slowly or do not decompose under normal environmental conditions. Others produce decomposition by-products that are even more hazardous than the original compound. Understanding the fate of PPCPs in the environment provides important information about the proper use and disposal of these products, and is an important step toward remediating the adverse effects of these materials.

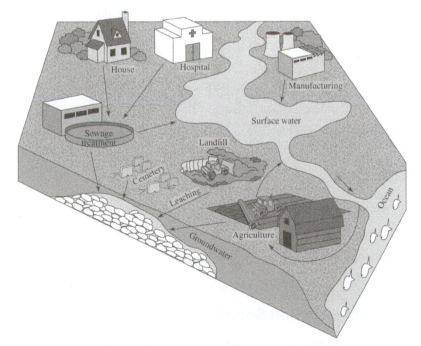

Figure 35.1. Some ways PPCPs can enter the environment.

In this investigation, you will explore the fate of caffeine under different conditions by measuring how fast it decomposes under normal and extreme environmental conditions. Here, caffeine will serve as a model drug, one that demonstrates the activity of other PPCPs found in the environment. Caffeine is appropriate for this study because it is one of the most commonly used drugs and is often found in the environment as a result of human use and disposal.

In chemistry, it is extremely important to know how long a reaction will take. For example, a drug molecule that takes 10 years to break down in the environment is different than a drug that takes 30 seconds. Chemical kinetics tell us about the **rate of reaction**, which is defined as how fast or slow a chemical reaction occurs. The rate of reaction is a measure of how the concentration of a reactant or product changes over time. In this investigation, you will study the rate of reaction for the degradation of caffeine under various pH conditions.

To determine a reaction rate, it is necessary to directly monitor the concentration of a reactant or product over time. For this investigation we will be using **UV-Visible spectroscopy**. Some molecules absorb ultraviolet (UV) or visible (Vis) light. The amount of light absorbed is related to the concentration of that molecule in solution:

$$\text{Absorbance (A)} = b \times \text{Concentration (C)}$$

Absorbance (A) is the amount of light absorbed by a sample, and b is an experimental constant. This equation shows us that the absorbance of light is correlated with concentration. Measuring the change in absorbance over time will allow you to determine the rate of reaction.

Making Predictions

After reading *Gathering Evidence*, prepare a table for recording your data. How do you think pH will affect the degradation of caffeine? Will the degradation reaction be faster or slower in acidic, basic, or neutral solutions?

Gathering Evidence

Overview of the Investigation

1. Use a pH meter to measure the pH of the reaction solutions.
2. Mix caffeine with each solution and monitor the reaction with UV-Vis spectroscopy.
3. Plot the data to determine reaction rates.

Part I. Measuring pH of Reaction Solutions

1. Obtain 20 mL of each reaction solution: 1-M HCl, acetate buffer with approximate pH 5, 9-M NaOH

CAUTION! Hydrochloric acid is corrosive and can cause burns. Avoid contact with skin, and always wear safety glasses. If you get any acid on your skin, immediately rinse with water and inform your instructor.

2. Carefully calibrate your pH meter. Operation of the pH meter is described in the *Laboratory Methods* section of this lab manual.

3. Measure and record the pH of your three reaction solutions.

Part II. Measuring Decomposition Rate of Caffeine

This part of the investigation is time sensitive and must be done quickly but carefully. Make sure that you have all equipment ready to use before mixing the caffeine with the reaction solution. Work with one reaction solution at a time, and carefully clean and dry your glassware before moving on to the next reaction.

1. Familiarize yourself with the operation of the UV-Vis spectrometer that you will be using. Set the measurement wavelength at 272 nm, and perform a blank measurement using a cuvette containing 1-M HCl.

2. Using a volumetric pipet, measure 3 mL of the caffeine solution into a 50-mL beaker, and then pipet 3 mL of the 1-M HCl reaction solution into a second 50-mL beaker.

3. Quickly but carefully, pour the contents of one beaker into the other and mix for 3 seconds. Then, transfer the mixture into a quartz cuvette so it is three-quarters full.

4. Place the cuvette into the sample holder of the spectrometer and immediately measure the absorbance of the sample. *Note:* If done correctly, this step should take no more than 10–15 seconds. You must record your initial absorbance reading as soon as possible after mixing the reagents.

5. Obtain absorbance measurements every 15 seconds for 5 minutes. Record the time after mixing and the absorbance for each measurement in your data table.

6. Dispose of your solutions in appropriate waste containers. Wash and dry the beakers and cuvette, then repeat the mixing and measurement with the acetate buffer.

7. Clean everything again and repeat the measurement process with 9-M NaOH.

Clean up

Dispose of all solutions in an appropriate waste container. Wash glassware with soap and water and allow it to drain.

Analyzing Evidence

1. Create a plot of absorbance vs. time for each of the three reaction mixtures. Time should appear on the *x*-axis of the graph, and absorbance should be plotted on the *y*-axis. Which reagent(s) caused a decrease in caffeine concentration over time?

2. For any reagents that did cause a change in concentration of caffeine, determine an average rate of decomposition between the initial measurement and 1 minute later. To do this, use the graphing program to calculate the slope of the decomposition line between these two times. Then, determine the average rate between 4 and 5 minutes. Are these rates the same?

Interpreting Evidence

1. For any reaction that showed a decomposition of caffeine, how does the rate of decomposition change with time? Does the reaction speed up or slow down as it proceeds? Explain your results.

2. The acetate buffer was chosen to mimic the natural pH of groundwater throughout the country. Use the U.S. EPA website to look up the pH of the groundwater in your area. Does the buffer adequately represent the conditions near you? Do you think caffeine that finds its way into your local environment will decompose quickly or slowly?

Making Claims

What can you claim about how pH affects the decomposition of caffeine?

Reflecting on the Investigation

1. Other investigations in this lab manual have asked you to measure the pH of water containing dissolved CO_2, and your textbook discusses this as well. Is carbonated water acidic or basic? On the basis of your answer and the results of this investigation, do you expect the caffeine in carbonated beverages to decompose or be stable?

2. Look up and draw the structure of caffeine and identify the functional groups in the compound.

3. The human digestive tract is acidic (the stomach contains approximately 0.16-M HCl). Under these conditions, will caffeine decompose quickly or slowly? Do you think the caffeine you consume in your morning coffee will completely break down in your body, or will some be excreted later in the day? Once you formulate a hypothesis, search for evidence to back it up. Cite your sources.

4. What are some consequences of caffeine contamination in the environment?

5. Will all drug molecules exhibit the same reactivity as caffeine? Why or why not?

6. Do a search for another PPCP that is found in the environment as a result of human activities. EPA's website can give you some ideas. Write a short paragraph explaining how the molecule enters the environment and the consequences of its presence to wildlife and the ecosystem as a whole.

Investigation adapted from Hopkins, T.A.; Samide, M. Using a Thematic Laboratory-Centered Curriculum to Teach General Chemistry. *J. Chem. Ed.* **2013**, *90*, 1162–1166.

NOTES

Synthesis of Mauve Dye

Asking Questions

- Define *synthetic* as it applies to chemistry.

- In what ways are synthetic dyes different from naturally derived dyes? In what ways are they the same?

- Who was William Perkin? Write down a few interesting facts about him.

- Why were people interested in synthesizing quinine?

- In this experiment you will be using household bleach, a 5.25% solution of sodium hypochlorite in water. Find the Safety Data Sheet (SDS) for this solution, and outline some of its hazards and the precautions you should take when working with it. Keep in mind that you will be using only a small amount (10 drops) in this experiment.

Preparing to Investigate

Imagine you are walking through London on a fall evening in 1855. What would you see? Under the glow of gaslights, fashionable people wear clothes dyed with hues taken from nature—brown, gray, yellow, and rusty red. London installed its first streetlights in 1807, fueled by coal gas. Coal gas is prepared by a gasification process involving heating coal with a little oxygen to break it down into more volatile components. Disposal of the gasification by-product, nonvolatile coal tar, became a problem. It could not be dumped in the river (the preferred method for industrial waste disposal at the time) because it was insoluble, and it would gum up the canals and kill the fish. When burned, it produced large amounts of black smoke.

Around the middle of the 19th century, chemists began to investigate using coal tar as a raw material for the synthesis of useful chemical products. In 1856, 18-year-old William Perkin, a student at the Royal College of Chemistry, embarked on a project to convert coal tar to quinine. Quinine, extracted from the bark of the cinchona tree, was important for fighting malaria in the distant tropical regions of the British Empire. Perkin never did manage to synthesize quinine, but he isolated a delightful purple dye that fixed on silk and was later named mauveine. Perkin had discovered the first aniline dye. He soon quit school and opened a dye works in Greenford, England, making a fortune supplying England with colorful cloth dyes.

Almost immediately, mauve became a very fashionable color, worn by Queen Victoria and Empress Eugenie, wife of Napoleon, as well as much of the general public. The British satirical magazine *Punch*, in a tongue-in-cheek article published in 1859, lamented the advent of a horrible disease, Mauve Measles, afflicting women in England. Symptoms included a "measly rash of ribbons," though affected men could be cured with "one good dose of ridicule." Related aniline dyes of many colors were developed and synthesized in England and elsewhere in Europe, and in less than a decade, Europeans could be found wearing clothing in a rainbow of bright colors– blue, green, purple and red—not available just a few short years before. A walk around London in 1865 would be a much more colorful affair!

Making Predictions

After reading *Gathering Evidence*, prepare a table to record your observations of the synthesis of mauve. Without using the word "mauve," describe the color do you expect from your dye.

Gathering Evidence

Overview of the Investigation
1. Prepare synthetic mauve dye.
2. Analyze the components of the dye using thin-layer chromatography.
3. Use the synthesized dye to color a piece of silk.

Part I. Synthesis of Mauve Dye
Note: The order of reagent addition is important to the success of your synthesis. Be sure to add all of the reagents in the order specified, and do not add more than one drop of *o*-toluidine or the dye you prepare will not be mauve-colored.

1. Into a plastic zippered bag, add 1 mL of white vinegar using a graduated plastic pipet.
2. Add 1 drop of *o*-toluidine into the vinegar. Mix by gently pressing on the bag with your fingers.
3. Next, measure 10 mL of ethanol into a graduated cylinder and add it into the bag.
4. Lastly, add 10 drops of bleach into the bag using a pipet. You should immediately see a color change, and the color will continue to evolve over the next hour. Record your observations.

CAUTION! Bleach is an oxidant that can damage skin and clothing. Wear gloves when working with it and take care to prevent it from dripping or spilling.

Part II. Thin-Layer Chromatography

1. If you are not familiar with the procedure for doing thin-layer chromatography (TLC), review it in the *Laboratory Methods* section.

2. Set up your TLC experiment as shown in Figure 0.9 in *Laboratory Methods*, using a piece of absorbent paper (such as filter paper) as the stationary phase, and a mixture of 75% white vinegar and 25% ethanol as the mobile phase. Make a concentrated spot of your dye about 1 cm from the edge of the plate, and make sure the spot is not submerged when you put it into the developing chamber.

3. Once the mobile phase is 1 cm from the top of the TLC plate, remove the filter paper. Analyze the TLC plate to determine if your synthesized dye consists of a single compound or a mixture of compounds. Your instructor may have you produce several chromatograms in order to monitor how the reaction mixture changes over time.

4. Draw your TLC plates on your data sheet, and dispose of the TLC plates in the chemical waste container.

Part III. Dyeing Silk

1. Place a small piece of white silk into your bag of dye, seal the bag, and gently push the silk around with your fingers so that it evenly absorbs the dye. After the silk has taken up the dye, carefully remove it from the bag and spread it on a paper towel to dry. Once the silk is dry, wash it with soap and water to remove any excess dye. Record your observations about the color-fastness of the dye in your notebook.

2. **Optional extension:** Your instructor may have available a small fabric test strip that you can use to dye several different types of fibers at once. Which fabrics are most effectively dyed by mauve? Which do not absorb much of the color?

STOP! Synthetic mauve dye will stain your clothing, so take care to prevent drips and promptly wipe up any spills with a paper towel.

Clean up

When finished, you may keep your dyed silk. Dispose of the TLC plates, plastic bags, pipets, and any excess dye in an appropriate waste container provided by your instructor.

Analyzing Evidence

1. What color is your dye? How did it change over the course of the reaction?

2. Using the formula given in the *Laboratory Methods* discussion of TLC, calculate the R_f for each spot on your filter paper. How many distinct spots are present?

Interpreting Evidence

1. How many different compounds are present in your dye? How do you know?

2. If you ran multiple TLC experiments over the course of the reaction, describe how the chromatograms change with time.

Making Claims

What can you claim about the identity of your dye? What color is it? Is it a single compound or multiple compounds? How effectively does it dye silk? Other fabrics?

Reflecting on the Investigation

1. William Perkin developed aniline dyes when he was experimenting with coal tar, a by-product of gas production. In what ways does the use of this material fit into the key ideas of green chemistry (see p. viii)?

2. Industrial production of aniline dyes began a surge of new development of chemical products, including pharmaceuticals, plastics, pesticides, and food additives. Identify and list five different synthetic chemical compounds or materials that you use every day.

3. Perkin's synthesis of mauveine used hexavalent chromium as the oxidizing agent, instead of bleach. You may have heard of this form of chromium if you have read a book or seen the movie about Erin Brockovich, who advocated for people affected by chromium in their drinking water in Hinkley, California. Explain how chromium is toxic (you may want to search for the SDS), and use the key ideas in green hemistry to describe how using bleach might be safer.

4. In 1888, *Punch* magazine published an ode to Coal Tar, the raw material from which aniline dyes and a host of other chemical products were made during the mid-19th century. The poem reads, in part,

> Oil, and ointment, and wax, and wine
> And the lovely colours called aniline;
> You can make *anything*, from a salve to a star,
> If you only know how to, from black Coal-tar.

Today, we use petroleum as our main raw material for the useful chemical products we use every day, from plastics to drugs. Chemists are also developing more sustainable feedstocks, such as fast-growing algae or switchgrass, from which to obtain our necessary products. Compose your own short poem as an ode to petroleum, or do some research into sustainable feedstocks and choose one of those as your subject.

5. In the late 1800s, people began noticing that workers in factories that produced aniline dyes suffered from unusually high rates of bladder cancer, and in 1915 Japanese scientist Katsusaburo Yamagiwa proved that coal tar could cause tumors. Today, most industrialized countries have workplace regulations that help keep workers safe. Do some research to find what laws in your country help to protect chemical workers. List the laws you find, with a short description of each and evaluate the effectiveness of each law.

6. Once the market for clothing dyes was established, manufacturers of aniline dyes began to include them in food and candy with little consideration of their effects on human health. Later, the dyes were found to be toxic, and regulatory bodies such as the U.S. Food and Drug Administration were created to regulate food additives. Imagine that you have developed a new food additive and wish to have it approved for use in your country. Do some research into what you would need to do, and list the steps you have to take and the agencies you need to deal with to have your additive approved.

7. The chemical structure of mauveine wasn't worked out until 1994, and it was found to be a mixture consisting mainly of the two compounds shown below, named mauvine A and mauvine B.

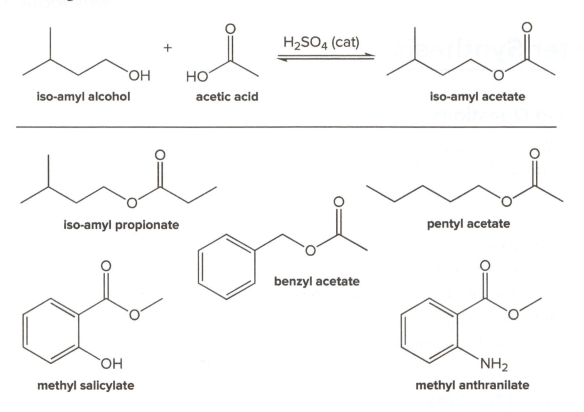

Figure 37.1. Synthesis of esters from alcohols and organic acids, and structures of the esters you will synthesize in this investigation.

Table 37.1. Esters to synthesize in this investigation.

Ester	Aroma	Reagents
methyl salicylate	wintergreen	methanol/salycilic acid
methyl anthranilate	grapes	methanol/anthranilic acid
benzyl acetate	peach	benzyl alcohol/acetic acid
pentyl acetate	apple/banana	pentanol/acetic acid
isoamyl acetate	banana	isoamyl alcohol/acetic acid
isoamyl propionate	banana/fruit	isoamyl alcohol/ propionic acid

Making Predictions

- Write the reactions that you will be conducting. Use Figure 37.1 as a model for what the reactions should look like.

- After reading *Gathering Evidence*, prepare a data sheet to record your synthesis plans and observations.

Gathering Evidence

Overview of the Investigation

1. Identify the reagents needed to synthesize the six esters.
2. Measure the alcohols and acids for the first three reactions.
3. Mix them, add the catalyst, and heat in a water bath.
4. When the reactions are complete, isolate the ester by pouring the reaction mixture into a concentrated sodium bicarbonate solution.
5. Repeat steps for the second set of reactions.

Preparing Esters

1. Determine what reagents are necessary to synthesize the following six esters: methyl salicylate, methyl anthranilate, iso-amyl acetate, iso-amyl propionate, pentyl acetate, and benzyl acetate.
2. Label small test tubes for the alcohols and acids for your first three reactions.
3. Measure 0.5 mL of each liquid reagent and 0.5 g of each solid reagent into labeled test tubes.

 CAUTION! Glacial acetic acid is extremely corrosive and should only be dispensed fume hood. Goggles and gloves are required when working with this substance.

4. Make a 70 °C water bath by heating water in a 250-mL beaker.
5. Combine the selected alcohols and acids in labeled test tubes, add 3 drops of concentrated sulfuric acid, and place the test tube in the water bath.

 CAUTION! Concentrated sulfuric acid is extremely corrosive and should only be dispensed in a fume hood. Goggles and gloves are required when working with this substance.

6. Heat for 10–15 minutes.
7. While the mixtures are heating, pour 20–25 mL of concentrated sodium bicarbonate solution into each of three 50-mL beakers.
8. After 10 minutes, check the reactions by wafting the fumes from each test tube toward your nose. If you smell the expected odor from the ester, quench the reaction by pouring the contents of the test tube into a beaker containing sodium bicarbonate solution. Confirm the odor of the ester by wafting the fumes towards your nose. The odor should

be obvious because excess acid is neutralized, and the volatile esters are not soluble in water, so they float on the surface. Record your observations on your data sheet.

9. If you do not smell the expected odor, return the test tube to the water bath so that the reaction can continue.

10. After the three reactions are complete, measure out the alcohols and acids for the next set of reactions.

11. Repeat Steps 5–9 for the second set of three reactions.

Clean up

Discard or recycle the contents of your beakers as directed by your instructor. Clean your glassware with soap and water.

Analyzing Evidence

1. Which of the esters had the most easily identified odor?
2. Did any of the syntheses take longer than the others?

Interpreting Evidence

1. Calculate the molar equivalents for one of the reactions. Which reagent was present in excess? How did that impact the equilibrium of the reaction?
2. How could you have shortened reaction times, especially for the reactions that took longer than the others?

Making Claims

What can you claim about esters and their synthesis?

Reflecting on the Investigation

1. Why was it easy to isolate the esters by pouring them into a basic solution? How did this process neutralize excess acidic reactants?
2. If you could have distilled the volatile ester product, how would this have set up a reaction that shifted the equilibrium towards the products?

Isolating DNA from Plant and Animal Cells

Asking Questions

- What property of DNA can be used to isolate it from cells?
- Why can the lysis and precipitation process be used to extract DNA from both plant and animal cells?
- How is DNA from different types of cells the same? How is it different?

Preparing to Investigate

With few exceptions, the cells of all living organisms contain a molecule called deoxyribonucleic acid (DNA), which provides the necessary instructions for growth and reproduction of that organism. Every person has unique DNA, so forensic scientists can use DNA analysis of samples related to a crime to help prove someone's guilt or innocence. In this investigation, you will use a standard procedure to isolate DNA from different samples. Although the procedure you will use in this investigation is very simple, it is similar to the way that hospitals and laboratories isolate DNA for medical and forensic applications.

DNA is a long polymer consisting of repeat units called **nucleotides** (see Chapter 13 of *Chemistry in Context*). Each nucleotide is composed of a phosphate group, a deoxyribose sugar, and one of four nitrogen-containing bases—adenine, cytosine, guanine, or thymine. The bases on two complementary strands form are attracted to one another and hydrogen bonding, which results in twisting into a double-helix shape.

Many methods have been developed for extracting DNA from cells. These procedures are designed to extract only the DNA and leave behind other cell components including, proteins and lipids. In general, DNA isolation methods begin with a **lysis** step, where the breakdown of cell walls is promoted by heat and a reaction with a detergent, a salt, and a pH-controlling reagent. Often an enzyme such as papain is added to break down the protein that encases the DNA. Once any remaining solids are removed by filtration, the DNA is **precipitated** from solution using an organic solvent such as ethanol or isopropanol. This process allows it to be isolated from other soluble components of the cell. DNA molecules precipitate in long strands that mat together and look something like coagulated egg white. The strands can be wrapped around a thin glass or wooden rod and removed from the liquid.

Making Predictions

- After reading *Gathering Evidence*, decide upon samples that you want to use to isolate DNA. Good choices include fruits and vegetables (fresh, canned, dried, or frozen) such as bananas, onions, peas, and strawberries; meats (both raw and cooked); and your own cheek cells that you can isolate by swishing 10 mL of pure water around your mouth and spitting it into a cup.

- Rank the samples you choose from highest to lowest in how much DNA will be isolated by the written procedures. Explain why you predict these results. For instance, will plant or animal samples produce more DNA? Will you expect a difference in DNA recovery between raw and cooked meat? Or between canned, frozen, and fresh vegetables?

- Prepare a data sheet that includes space for predictions, observations, and any procedural changes.

Gathering Evidence

Overview of the Investigation

1. Obtain and weigh your samples.
2. Homogenize and heat your sample. Use enzymes to break up the proteins if necessary.
3. Chill the mixture and filter it.
4. Precipitate the DNA with alcohol.
5. Spool the DNA around a glass or wooden rod.

Extracting and Isolating DNA

A. Preparation of homogenizing solution

1. Chill a bottle of isopropyl or ethyl alcohol by placing it inside a bucket or other large container and surrounding it with ice.

2. Make homogenizing solution by combining 10 mL of Woolite® detergent, 2.5 g of NaCl, and 90 mL of water in a blender and blending at high speed until mixed.

3. Pour the blended solution into a beaker, and place the beaker into a hot water bath (at about 60 °C) for approximately 15 minutes. After the solution has warmed, carefully remove the beaker from the hot water bath and pour the solution back into the blender.

B. Preparation and lysis of sample

The larger your sample, the more DNA you will be able to extract. For instance, if you are using dried or frozen peas, try measuring out 150 mL of them. Half of an onion or a whole banana work well. However, for some samples (like cheek cells) you will not be able to have a large sample. The amounts listed here are guidelines. If at first the extraction doesn't work, try again with more sample (if available) or more homogenizing solution.

1. Weigh your sample and record its mass.
2. Mash your sample with the homogenizing solution. Use about 20 mL of solution for every 10 g of sample. You can do this by hand, or use the blender to mix the sample and the solution.
3. Filter the solution through two layers of cheesecloth placed into a wide-mouth funnel.
4. To the filtered liquid, add a small spatula scoop of papain enzyme or meat tenderizer, or 2–3 drops of pineapple juice or contact lens cleaning solution. This is an optional step for plant-based samples, but necessary for animal samples. The purpose of this step is to break up any protein bound to the DNA, allowing it to be more easily isolated later.
5. Place the solution into an ice bath and allow it to settle for 5–10 minutes.

C. Precipitation of DNA

1. Fill a test tube about one-third full with the filtrate solution. Use a large test tube if you have a large sample, or a small test tube for a small sample. You do not need to use all of the filtrate if you have a lot of it.
2. Tilt the tube and carefully pour about the same volume (or a little more) of cold alcohol down the edge of the tube so that it forms a layer on top of the filtrate. Try to prevent the two liquids from mixing. Allow the tube to sit undisturbed for at least 5 minutes. Carefully observe the precipitation of DNA, which will collect at the interface between the two liquids.
3. Gently place a clean glass stirring rod or a wooden rod (such as a long toothpick or chopstick) into the tube and slowly turn it to collect the DNA. Sometimes, the DNA will wrap around the surface of the rod and can be removed from the test tube.
4. Record your observations about the amount of DNA collected from each sample, and what it looks like. If you were able to recover some DNA from the tube, measure and record its mass.

D. Analysis of DNA

Depending on the equipment available in your laboratory, you may be able to further analyze your DNA. For example, you can put a small amount of your collected DNA on a microscope slide, stain it with methylene blue, and look at the strands under a microscope. Alternatively, your instructor may demonstrate or allow you to separate your DNA using a technique called gel electrophoresis. This separates the pieces of DNA you collected based on their molecular size.

Analyzing Evidence

1. Describe the appearance, amount, and physical properties of the DNA you isolate.
2. Rank your samples based on the amount of DNA that you isolated. Did it match your predictions?

Interpreting Evidence

1. DNA was precipitated out of a water solution by adding isopropyl or ethyl alcohol. Why do you think DNA is less soluble in ethanol than in water? Remember that these alcohols are partly hydrocarbon and therefore less polar molecule than water.
2. Do some research into other tests could you could do on your DNA. How do forensic scientists analyze DNA?

Making Claims

What can you claim about the isolation of DNA from plant and animal cells? What evidence from this investigation supports your claims?

Reflecting on the Investigation

1. Below are the structural formulas of the four nucleotide bases in DNA: adenine, cytosine, guanine, and thymine. Label each one and name three features that they have in common.

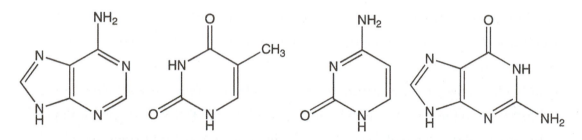

2. Which hydrogen atoms in the four nucleotides are responsible for forming the hydrogen bonds that hold DNA strands together in the double helix? How many hydrogen bonds are possible for each? See Chapter 13 of *Chemistry in Context* for help. Suggest an explanation for why adenine always pairs with thymine and cytosine always pairs with guanine.
3. In light of your answer to Question 2, suggest why DNA molecules that are rich in cytosine and guanine will decompose at a higher temperature than DNA rich in adenine and thymine.
4. The diversity of life depends on the fact that a great many combinations of the four nucleotide bases can be assembled into long chains. Illustrate this by writing out 20 of the 64 possible combinations of just three bases, using any combinations of the four building blocks (designated as A, T, C, and G).

Chromatographic Study of Dyes and Inks

Asking Questions

- How can you determine whether a substance is pure or a mixture?
- What properties can be used to separate components in a mixture?
- What aspect of the inks in "washable" markers would enable them to be separated under the conditions of this investigation?
- How might a forensic scientist detect alterations to a document?

Preparing to Investigate

How can we know if something is pure, containing just one element or compound, or a mixture of several substances? Chapter 1 in *Chemistry in Context* introduces the concepts of pure substances and mixtures, and notes that most of the materials we encounter daily, such as air, water, and food, are mixtures of chemical substances.

In this investigation, you will study inks from washable felt-tip pens to find out whether the colors are produced by a pure substance or a mixture of substances. By careful comparison, it should be possible to determine whether some of the substances have colored components in common. Additionally, you may be given the task of matching an unknown ink to a known pen, something that forensic scientists are asked to do. You may also investigate what happens when someone tries to alter a check or another document by dissolving the ink with a solvent or removing the color with household bleach, and how such alterations may be detected.

The ink investigation uses **chromatography**, a common method for separating and studying the components of a mixture. Chromatography is described in detail in the *Laboratory Methods* section, which you should read before doing this investigation. In chromatography, the components of a mixture are allowed to move along a stationary surface. Each component in a mixture retains its own properties and moves at a rate determined by its own characteristics. Since they move at different rates, the components become spread out and separated from each other like runners in a race. When paper is the stationary surface, the usual strategy is to place small spots of substances near one edge of the paper. That edge is placed in a liquid, such as water, and the liquid moves up the paper by capillary action. The liquid carries the components with it, but at differing rates depending on their solubility.

Making Predictions

- Suggest some reasons why some components of the dyes will move up the filter paper at different rates.
- After reading *Gathering Evidence*, prepare a data sheet to record your observations. Be sure to leave room for the paper chromatography strips and to identify the solvents.

Gathering Evidence

Overview of the Investigation

1. Use chromatography to analyze the inks in water-soluble pens.
2. Identify the ink used to make an unknown document.
3. Investigate the effects of various solvents and bleach on inks and paper in documents.

Part I. Separating the Components of Inks

These investigations should be done individually so that each person has a completed chromatogram, but you are encouraged to compare results and discuss the interpretation.

1. Obtain four pens that contain different colors of vivid, washable inks. Good choices are the Vis-à-Vis overhead-projector pens, LiquidMark Washable Markers, or any other markers labeled washable.
2. Obtain a rectangular piece of filter paper, approximately 5 cm x 12 cm. Use a *pencil* to write your name at the top and to draw a line 2 cm from the bottom edge of the paper (Figure 39.1). Make four small pencil marks, 1-cm apart, along the line.

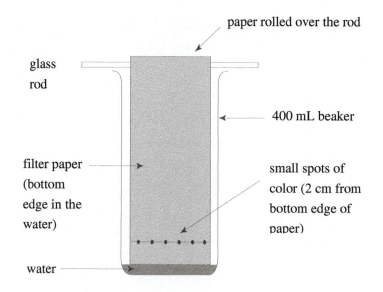

Figure 39.1. Chromatography setup.

3. On each pencil mark, make a small dot using a water-soluble pen. Use a different pen for each spot, and label the color and/or brand of the pen in pencil near each mark.

4. Lay a glass rod across the top of a 400-mL beaker. Fold or roll the top edge of the filter paper so that it hangs on the rod inside the beaker with the bottom of the filter paper close to, but not touching, the bottom of the beaker.

5. Carefully, add a small amount of water so that there is a shallow pool of pure water in the beaker. The bottom of the paper should extend into the water, but the water level should not be higher than the line that contains the spots. Leave the beaker undisturbed on the desk and observe what happens.

6. When the top edge of the water has moved about two-thirds of the way up the paper, remove the paper and lay it on a paper towel to dry. While the paper is still damp, draw a line *in pencil* to show the top edge of the wet area.

7. Replace the water with another solvent, such as methyl alcohol, ethyl alcohol, or rubbing alcohol (which is a mixture of isopropyl alcohol and water). Do the investigation the same way, and record your observations.

8. Obtain a document from your instructor that has been produced with one of the pens. Devise a method for extracting the ink from the document and analyzing it using chromatography. Can you determine which pen was used to make the document?

Part II. Investigation of Altered Documents

1. Divide a voided check or a blank piece of paper into four to six portions by drawing lines with a pencil. Write within each section with a one of the pens you analyzed in Part I.

2. Choose solvents to use to try to remove the ink from the paper. Some solvent possibilities are: water, acetone, ethanol, xylenes (or another hydrocarbon solvent), and household bleach (~5% sodium hypochlorite solution in water). Label each section on your paper *in pencil* with the names of the solvents you will use to try to remove the ink.

3. Try to remove the ink from the paper using the following procedure: dip a cotton swab or cotton ball into the solvent, and gently wipe it on the paper over the ink. Do not use more than one solvent in the same section of the paper. Record your observations about what happens to both the ink and the paper with each solvent.

4. Look at the paper under a UV light and record your observations.

5. Put the paper into an iodine chamber and let it sit for a few minutes to interact with the iodine vapor. What do you observe? Remove the paper from the iodine chamber and allow it to sit for a few minutes until the color fades somewhat.

6. Use a cotton swab to spread a thin layer of the KI-glycerol reagent over the paper. Record your observations.

7. **Optional extension:** Would any of these solvents make a good invisible ink? Think about which ones don't alter the appearance of the plain paper, but can later be detected in some way. With your instructor's permission, write your partner a message by dipping a cotton swab into a solvent and writing on a piece of paper. See if your partner can use one of the visualization methods described here to see and decode your message.

Analyzing Evidence

1. When the filter papers are completely dry, staple or tape each one to your data sheet. Include a brief summary of what samples and solvent were used.

2. Using the formula given in the *Laboratory Methods* discussion of thin-layer chromatography, calculate the R_f for each spot on your filter paper. How do the R_f values compare? Can you identify any spots that have the same R_f values?

3. Look at your data. Determine which inks contain a single colored substance and which are mixtures. Describe your evidence.

4. Look at your data and determine which samples have colored components in common with other samples. Explain your reasoning.

5. If you used a check for Part II, what visual changes did you observe for the paper with each solvent?

6. Which (if any) of the solvents were effective at removing the ink from the paper? Which solvents used to remove the ink on the check or paper could be most easily detected?

Interpreting Evidence

1. From your results, predict which colored components are most soluble in water and which are least soluble in water. Explain your reasoning with specific examples from your data.

2. Which of your samples had the most number of components? Explain any trends in the data that you see.

3. What would be the most effective method to remove ink from the paper? What would be the most effective method to detect the alteration?

Making Claims

What can you claim about the similarities and differences in the dyes and inks that you sampled?

Reflecting on the Investigation

1. Why was it important to use a pencil to label the filter paper?
2. Suggest a possible explanation for why the chromatography results using alcohol are different from those in water.
3. If you used a check in Part II, you saw security features on the documents to detect forgeries. How do these features work?

NOTES

Drug Money: Detection of Drug Residues on Currency

Asking Questions

- What are some ways that residues of illegal drugs such as cocaine and methamphetamine come to be present in paper currency?
- What type of bills (age, denomination) do you think will have the highest concentration of drug residues?
- How do you think drug residues on currency collected from different areas of the country or the world will differ for each drug? Consider both regional differences as well as urban compared to rural areas.
- The reagent you will use in this experiment contains concentrated sulfuric acid. Look up the Safety Data Sheet (SDS) for this substance online, list its hazards, and explain what you should wear and do during the experiment to keep you safe while working with it.

Preparing to Investigate

If law enforcement officers find an unknown substance that they suspect to be an illegal or controlled substance, they usually send it to a forensic laboratory for analysis. Unambiguous identification of an unknown substance can take significant time and utilizes expensive instrumentation such as mass spectrometry, so forensic scientists often perform **presumptive drug tests** that qualitatively suggest the presence of drugs. The results of these tests are not admissible in court because many tests give a positive result for both controlled substances and over-the-counter drugs. However, they are often used to screen samples to determine which are worth further analyzing to gather admissible evidence. These presumptive tests are quick and usually involve a chemical reaction that results in a color change, possibly indicating the presence or absence of a drug.

Some of the currency circulating in the United States is contaminated with residues of illegal drugs, such as cocaine and methamphetamine, which are often used by snorting the powdered drug through rolled up bills. The residues of these drugs can be detected using presumptive drug tests. The National Institute of Justice in the United States has [?] is an battery of presumptive tests that give different results for different compou[?] one of these tests, the Scott reagent, available to use in your investig[?]

aqueous solution of cobalt thiocyanate (Co(SCN)$_2$). It will give a positive test for cocaine as well as several over-the-counter drugs by turning from pink to greenish-blue.

Making Predictions

Devise a series of questions that you hope to answer by testing currency for drug residues. You may wish to do some research before coming to the lab to find out how other researchers have observed the ways drug residues vary on currency from different locations and of different denominations or ages. Your instructor may have you work together as a class to ask and answer questions about drug residues on currency by combining your class data. Determine how much currency and what type you'll need to answer your questions. Could you obtain bills from places outside your city or town to test regional variation in drug residues? Or perhaps investigate whether older or newer bills have more drug residue? Or bills of different denominations? Gather some paper currency from your home, local stores, or a bank. The currency will *not* be destroyed and can still be spent after doing the experiment, so feel free to bring in bills of higher denominations if you wish.

Gathering Evidence

Overview of the Experiment

1. Test known and unknown drug substances using presumptive drug tests.
2. Use the Scott reagent to test known contaminated bills and circulated currency.

 STOP! Concentrated sulfuric acid is corrosive and can be hazardous! Wear eye protection, a lab coat or apron, and gloves at all times in the laboratory. If you get the reagents on your skin or in your eyes, immediately rinse with copious amounts of cold water for several minutes, notify your instructor, and seek medical attention if necessary.

Part I. Testing Drugs and Unknown Substances Using Presumptive Tests

Presumptive drug tests can be used to detect a number of controlled and over-the-counter substances, including aspirin, acetaminophen (active ingredient of Tylenol®), pseudoephedrine HCl (Sudafed®), brompheniramine maleate (Dimetapp®), diphenylhydramine (Benadryl®), opiates extracted from poppy seeds, table salt, and sugar. Your instructor may provide you with known and unknown samples to identify. To perform the tests, follow this protocol:

Put a small amount of the substance in a well of a wellplate. If the substance is a solid, the tip of a small spatula to add approximately 5–10 mg. If the substance is aqueous, 3 drops to a well using a pipet.

2. Add 1–2 drops of the reagent to your sample, and stir gently with a toothpick. Note the time that you added the reagent, and observe the color change over the next minute.

Part II. Detection of Residues on Currency

Not all currency is contaminated with illegal drugs, so in order for you to see how the test works your instructor will provide you with bills that have been deliberately contaminated with an over-the-counter drug that gives positive presumptive test results. Be sure to return the bills to your instructor after this part of the experiment so that they can be reused. The procedure is the same whether you are using deliberately contaminated bills or circulated currency that you brought from home.

1. Use a moist (but not too wet) cotton ball, cotton swab, or small piece of paper towel to wipe both sides of the bill. Use the same part of the wipe for both sides of the bill to maximize the concentration of any drug that may be present. If you have several bills available, you may want to swab two or three bills to ensure that any residues are detected. Any cocaine that is present on the bill is readily soluble in water, but to maximize transfer of the residue to the cotton the swab should be damp but not too wet.

2. Place the absorbent swab on a clean surface with the section that wiped the bills facing up. Place a few drops of the Scott reagent on the surface, and look for a color change. Record your observations.

Part III. Optional Further Experiments

*   Other presumptive drug tests can be used to look for various over-the-counter and illegal drugs. Some that your instructor may have available are the Mandelin, Marquis, and Mecke reagents, which give positive results for methamphetamine and some other drugs. If these other tests are available, you can use them to test known and unknown samples.

*   Another presumptive drug test, the Duquenois-Levine reagent, consists of a solution of acetaldehyde and vanillin in ethanol. This reagent will indicate the presence of tetrahydrocannibinol (THC), the active ingredient in marijuana, by a change to a blue or purple color. In states or countries where possession and use of marijuana is legal, your instructor may provide you with this reagent and samples to test.

Analyzing Evidence

1. List the substances that gave a positive Scott reagent test result in Part I.

2. Describe in detail the changes you observed using the Scott reagent test on the known contaminated bills. Did you see the same changes on any of the bills you brought from home?

Interpreting Evidence

1. Would a positive Scott reagent test allow you to identify a specific drug that tests positive? If you tried other reagents, answer the same question for those tests.
2. Did you see any differences in the currency samples you tested? Did any bills have no indicated drug residue? Did any have a lot? How do your results match your hypothesis?

Making Claims

What can you claim about the presence or absence of drug residues on the currency that you tested? Can you draw any conclusions about regional or denominational variation in the drug residues? Do you see any trends?

Reflecting on the Investigation

1. How successfully did your experiment answer the question you asked in the *Making Predictions* section? If you were to do this investigation again, how would your experimental design change?
2. For any of your bills that gave a positive presumptive test, have you *proved* the presence of a controlled substance? Why or why not?
3. Explain, in your own words, why presumptive tests like those you did here are not admissible as evidence in court. Outline some of the problems with the tests, and then explain why they are still useful to forensic scientists as preliminary indicators of the presence of drugs in a sample.
4. The following are two methods that forensic scientists use to further analyze samples that have positive presumptive tests, in order to provide identification that is admissible in court. Do some research about each technique; describe how it works and how long it might take to identify a compound using the following methods:
   a. Mass spectrometry
   b. High-performance liquid chromatography

Investigation adapted from Thompson, R. B.; Thompson, B. F. *Illustrated Guide to Home Forensic Science Experiments,* 1st Edition; DIY Science; Maker Media, Inc: San Francisco, CA, 2012.

Glossary

Absorbance (A) – the portion of light that gets absorbed (or does not get transmitted) by a sample in spectroscopy

Acid – a chemical that donates H^+ to water

Acidity – the amount or concentration of acidic substances present in a solution; usually measured by titration with base or calculated from pH

Alkalinity – the amount or concentration of basic substances present in a solution; usually measured by titration with acid

Amino acid – a compound consisting of a carbon atom to which a carboxylic acid, and amine group, a hydrogen atom, and side-chain group "R" are attached; amino acids are the monomers that link together to make proteins

Barometer – an instrument used to measure atmospheric pressure

Base – a chemical that accepts H^+ from water, or donates OH^- to water

Battery – a collection of galvanic cells wired together

Biodiesel – a biofuel prepared through a transesterification of triglycerides with a small alcohol

Biofuel – energy source derived from a biological, usually renewable, source rather than from petroleum

Blank sample – a sample containing only solvent or air, used to determine baseline absorbance or transmittance in spectroscopy

Calorimetry – a method for measuring the heat released by a substance upon combustion or other chemical reaction

Carcinogen – a substance that is capable of causing cancer

Casein – the primary protein found in cow's milk

Chromatography – a method for separating compounds based on their differing affinities for a stationary phase and a mobile phase

Combustion – the reaction of a fuel with oxygen, that results in the release of energy

Critical point – the temperature and pressure beyond which a substance does not have distinct liquid and gas phases

Cross-linked polymers – rigid polymers in which the polymer chains are covalently linked together

Dehydration – a chemical reaction that results in the loss of water

Density – the mass of a substance divided by its volume

Distillation – a separation process in which a liquid solution is heated and the vapors are condensed and collected

Electrolytic cell – a type of electrochemical cell in which electric energy is converted into chemical energy; it can be considered the opposite of a galvanic cell

Electromagnetic spectrum – the range of all possible frequencies of electromagnetic radiation

Esterification – a chemical reaction that produces an ester functional group, usually by reaction of a carboxylic acid with an alcohol

Extraction – the use of a solvent to separate soluble compounds in a mixture from insoluble compounds

Extrapolation – estimating values beyond the measured data on a graph

Fermentation – a process that uses yeast to break down sugars into ethanol and carbon dioxide

Fume hood – a laboratory apparatus that draws air and toxic fumes away from the chemical researcher

Galvanic cell – an electrochemical cell that converts the energy released in a spontaneous chemical reaction into electrical energy; it is the opposite of an electrolytic cell

Gravity filtration – a process to separate a liquid and a solid where the solid collects in a filter while the liquid travels through the filter unassisted except by gravity; used when a scientist wishes to collect and retain the liquid

Green chemistry – a way of doing chemistry that seeks to protect the health of humans, wildlife and the wider environment.

Hydrocarbons – compounds that contain only hydrogen and carbon

Hydrophilic – a compound or substance that readily dissolves in water

Hydrophobic – a compound or substance that does not dissolve in water

Green chemistry – a philosophy that encourages the reduction of waste and the use of less toxic compounds and processes in chemical synthesis and engineering

Indicator – a chemical substance that undergoes a visible change upon reaction

Interpolation – estimating values between measured points on a graph

Ion – an atom or group of bonded atoms bearing a positive or negative charge

Lysis – breakdown of cell walls through a chemical reaction

Mass – The amount of matter in a sample

Meniscus – the curved surface of a liquid

Miscible – describes two liquids that can freely mix in any proportion

Mobile phase – in chromatography, the part of the system that moves over the stationary phase; usually a liquid or a gas

Molar mass – the mass, in grams, that contains 6.02×10^{23} (Avogadro's number) of atoms of an element, or molecules of a compound

Molarity (M) – the number of moles of a substance in exactly 1 liter of solution

Mole – the amount of a substance containing 6.02×10^{23} atoms or molecules

Monomer – a small molecule used to prepare a polymer, or the repeat unit of a polymer

Nanometer (nm) – a unit of distance often used for the wavelength of visible light, equal to 1×10^{9} meters

Natural polymer – see *polymer*

Nucleotide – repeat unit of DNA consisting of a phosphate group, deoxyribose sugar, and nitrogen-containing base

Octet rule – states that atoms tend to share enough electrons with other atoms that they will have eight electrons (an octet) in their outer valence shell

Percent transmittance (%T) – the amount of light that passes unaffected through a sample in spectroscopy; the light that is not absorbed

pH – a measurement of the acidity or basicity of a solution

Plastics – polymers that can be molded into a shape while warm and then set into a more rigid form

Polymer – a long molecular chain composed of smaller repeating units; *natural polymers* occur in biological systems, while *synthetic polymers* are prepared in the laboratory

Precipitate – removal of a soluble molecule from solution by addition of a different solvent or a salt that causes the molecule to become insoluble

Pressure – the amount of force exerted by a substance on its surroundings.

Presumptive drug test – a qualitative test that can detect the presence of a drug

Primary protein structure – the order of the amino acids linked together in a protein; it can be changed by breaking or forming amide bonds between the amino acids

Protein – a naturally occurring polymer composed of amino acids linked together by amide bonds

Rate of reaction – how fast or slow a chemical reaction occurs, or how the concentration of a reactant or product changes over time

Regression – a best-fit straight line for a set of data plotted on a graph

R_f – short for "retention factor" or "retardation factor"; in thin-layer chromatography, R_f is the distance traveled by a compound up the plate, divided by the distance traveled by the solvent

Slope – the steepness of a linear portion of a graph, given the rise divided by the run

Solder – a low-melting mixture of several metals that is melted onto wires to make an electrical connection

Solubility – the maximum mass of solid that can dissolve in a given volume of water

Specific heat – the amount of energy required to raise the temperature of 1 gram of a substance by 1 °C

Spectrophotometer – a specialized laboratory instrument that measures the absorption of light by matter

Spectroscopy – the use of light to study matter

Spectrum – a graph depicting absorbance or % transmittance of a sample at different wavelengths of light

Stationary phase – in chromatography, the part of the system that doesn't move; usually a solid

Sublimation – a phase change in which a substance transforms from a solid to a gas

Supercritical fluid – a phase existing beyond the critical point, when a substance exhibits characteristics of both liquid and gas

Synthetic polymer – see *polymer*

Temperature – the measure of the average kinetic energy of particles in a sample; in science, it's measured in $^{\circ}C$

Tertiary protein structure – the overall three-dimensional shape of a protein that results from interactions between the amino acid side chain groups

Thermometer – the instrument used to measure temperature

Thermosetting plastic – a polymer material that does not melt

Thin-layer chromatography (TLC) – a type of chromatography involving a stationary phase consisting of a glass, plastic or aluminum plate coated with solid silica and alumina, and a mobile phase consisting of a liquid solvent

Titration – a method of analysis that uses a chemical reaction to determine the concentration of a substance; an indicator is usually used to provide a visual cue that a reaction has occurred

Total dissolved solids (TDSs) – the amount of dissolved material (usually ionic salts) present in a sample of water

Total hardness of water – a measure of the concentration of calcium and magnesium ions in a water sample

% Transmittance (%T) – in spectroscopy, the amount of light that passes through a sample, rather than being absorbed

Triglyceride – a component of fats and oils consisting of three fatty acids attached to one glycerol molecule by ester linkages

Triple point – the temperature and pressure at which a substance exists in three phases: solid, liquid and gas

Valence electrons – the electrons in the outermost, or valence, shell of an atom

Viscosity – the ability of a liquid to resist flow; *viscous* liquids are thick and syrupy

Volume – the amount of space taken up by a sample; the volume of liquids are usually measured using a graduated cylinder

Wavelength – the length of one wave of electromagnetic radiation; light of longer wavelength is lower in energy

Wet chemistry – chemical reactions or analyses performed in the liquid phase

***y*-intercept** – the value of a linear portion of a graph where the line crosses the *y*-axis, or where $x = 0$

NOTES